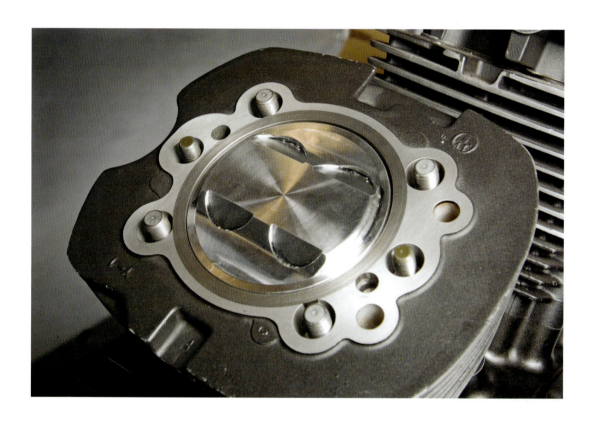

How To Modify Your
Metric Cruiser

Evans Brasfield

MOTORBOOKS

Dedication:
For Minna, my guiding light.

First published in 2005 by Motorbooks, an imprint of MBI Publishing Company, Galtier Plaza, Suite 200, 380 Jackson Street, St. Paul, MN 55101-3885 USA

© Evans Brasfield, 2005

All rights reserved. With the exception of quoting brief passages for the purposes of review, no part of this publication may be reproduced without prior written permission from the Publisher.

The information in this book is true and complete to the best of our knowledge. All recommendations are made without any guarantee on the part of the author or Publisher, who also disclaim any liability incurred in connection with the use of this data or specific details.

This publication has been prepared solely by MBI Publishing Company and is not approved or licensed by any other entity. We recognize that some words, model names, and designations mentioned herein are the property of the trademark holder. We use them for identification purposes only. This is not an official publication.

Motorbooks titles are also available at discounts in bulk quantity for industrial or sales-promotional use. For details write to Special Sales Manager at MBI Publishing Company, Galtier Plaza, Suite 200, 380 Jackson Street, St. Paul, MN 55101-3885 USA.

ISBN-10 0-7603-2142-6
ISBN-13 978-0-7603-2142-3

Editor: Peter Schletty
Designer: Sara Grindle

Printed in China

On the front cover: 2005 Victory Hammer

On the frontispiece: Notice how the flat of the piston doesn't rise above the top of the cyclinder. If it did, it would contact the head when the was running—a very bad thing.

On the back cover: Installing the headers into the muffler requires some finesse to get everything lined up.

About the author:
Evans Brasfield is a writer and journalist specializing in motorcycles. Prior to setting out on his freelance career, he was Feature Editor at both *Motorcycle Cruiser* and *Sport Rider* magazines. He supported himself in graduate school by teaching motorcycle safety for the California Motorcycle Safety Program. In pursuit of his passion for two wheels, he has ridden everywhere he can (including all the way up to the Arctic Ocean) and gets a perverse pleasure out of riding in the rain. He feels that riding cruisers, particularly touring, is about the most elemental motorcycling experience around—keeping him in touch with the true roots of motorcycling.

Evans lives in Burbank, California, with his wife, Karin, and their daughter, Minna.

Contents

Acknowledgments 6

Section One: Basics 7

Section Two: General Maintenance 14
    Project 1:   Adjusting Clutch and Throttle Freeplay 14
    Project 2:   General Lubrication 18
    Project 3:   Oil Change 21
    Project 4:   Coolant Check and Change 26
    Project 5:   Cleaning/Replacing Air Filter 28
    Project 6:   Spark Plug Check and Replacement 30

Section Three: Brakes 34
    Project 7:   Brake Pad Change 34
    Project 8:   Drum Brake Maintenance 39
    Project 9:   Hydralic Fluid Change 43
    Project 10: Stainless-Steel Brake Line Installation 47
    Project 11: Caliper and Master Cylinder Rebuild 50
    Project 12: Aftermarket Caliper and Disc Installation 55
    Project 13: Installing Aftermarket Wheels 58

Section Four: Suspension 64
    Project 14: Adjusting Preload 64
    Project 15: Installing Fork Springs 68
    Project 16: Changing Fork Oil 71
    Project 17: Replacing Fork Seals 74
    Project 18: Installing Cartridge Emulators 77
    Project 19: Installing an Aftermarket Shock 83
    Project 20: Lowering a Cruiser 86

Section Five: Engine 90
    Project 21: Installing a Full Exhaust System 90
    Project 22: Synchronizing Carburetors and EFI Throttle Bodies ..96
    Project 23: Installing a Jet Kit 99
    Project 24: Installing an Aftermarket Carburetor on a V-Twin ..102
    Project 25: EFI Tuning 104
    Project 26: Dyno Tuning CV Carburetors 107
    Project 27: Installing an Aftermarket Air Cleaner 110
    Project 28: Installing a Dyna 3000 Ignition 112
    Project 29: Clutch Replacement 114
    Project 30: Installing a Big Bore Kit 118

Resources 125
Index 126

Acknowledgments

As this, my second tome for Motorbooks, draws to a close, I've noticed that many of the same people who I thanked in my first one are here again. My editor, Darwin Holmstrom, dealt with many phone calls interrupting his busy schedule. Peter Schletty (poor guy) was tasked with keeping me focused and as on deadline as I'm capable of being. Pat Hahn, again, read the text and offered perceptive comments, improving the book immensely. Andrew Trevitt read the text before anyone else, pointing out inaccuracies. Without all you people, this project would not have happened.

Producing a how-to book would have been impossible without the wholehearted support of the motorcycle manufacturers. Honda, Kawasaki, Victory, and Yamaha all were generous enough to loan me bikes to use for photography. I can't thank these companies enough for entrusting their machinery to me. One person, however, stood head and shoulders above the rest when I was up against my deadline. Honda's Jon Siedel came through with bikes when I was in dire need. Thank you.

Although I know I'm stating the obvious by saying that the aftermarket is, in many ways, the heart and soul of the cruiser market. The talented folks who dream up, design, test, and manufacture the parts that we grace our cruisers with are not only the driving force behind this book, but also the amazing growth of cruiser sales. Yes, the OEMs make cool bikes, but cruiser riders tend to modify their bikes more than any other riders. The aftermarket makes it possible. Take a look at the resources in the back of the book for a complete list of those involved. As always, a few individuals deserve a special mention: Nigel Patrick, Chris Taylor, Scott Valentine, John Vaughan-Chaldy, and David Zemla. Jamie Elvidge at *Motorcycle Cruiser* also deserves thanks for not just being my friend and frequent employer. She also refrained from calling the police when I stole her Mean Streak in my mad rush to finish the book's photography.

Finally, I want to thank my wife, Karin Rainey. Aside from bringing home the bacon, she also happens to be a great mother and the love of my life. Mere words cannot express how lucky I feel to have found you.

Evans Brasfield
www.cruiserprojects.com
September 2005

SECTION 1
BASICS

Deciding to perform modifications to your cruiser on your own is the first step in a long, fascinating journey. As with any endeavor, you need to plan the initial portion of your route. From the most basic perspective, you need a place to work on your bike, a factory service manual to guide you through the intricacies of the modern motorcycle, and the right tools—including special-use ones—to do the actual work. With the right attitude, you'll soon find yourself comfortable with just about any excursion into your motorcycle's inner workings.

Setting Up Shop

The days of shade-tree mechanics are over. The advancement of motorcycle technology killed them off long ago—though some claim they still do exist. The truth is that you need a dedicated space to work on your bike if you really want to do more than maintenance. The importance of having someplace indoors should be obvious, especially for a novice mechanic. You're going to get stuck and need to go out for more parts, advice, cold beer, or all of the above. Also, some projects require multiple days. You don't want your engine internals to be open to the elements, do you?

Your garage doesn't have to be the Taj Mahal—although we can always dream. At the bare minimum, you need a cement floor (clean and level are nice extras). Electricity and good light are vital. A work bench is also good, but not necessary if you're just starting out.

Most important, you need a secure place where you can safely leave a myriad of parts out in an arrangement that works for you as you wrench. (At my last apartment, I had a neighbor's pot-bellied pig walk up to my bike and actually eat the exhaust manifold nuts I had set in front of the bike while I reinstalled the system.)

As you acquire more and better tools, you'll want to leave them safe and organized in your shop, too. Invest in some kind of toolbox, chest, or cabinet to store your tools.

Factory Service Manual

As you read through this book, you'll see constant references to the factory service manual. Every bike is different, and while I've tried to outline the bulk of the information you'll need to perform the projects here in a universal way, you'd be foolish to attempt many of them without your bike's factory service manual. Not only does the service manual give you important tips on how to access parts on your particular motorcycle, but it also lists all the specific measurements (gap, length, runout, thickness, etc.) as well as torque figures used for reassembly.

A factory service manual will itemize any special tools you need to perform specific maintenance chores on your bike. Since original equipment (OE) tools are usually quite expensive, you can use the manual to see if you can find a third-party tool that will perform the same function at a considerable savings.

Finally, a manual becomes a vital research tool as you move into more advanced projects. It will give you a complete idea of what the project will entail, how long it will take, the tools you'll need, and the parts involved. After working on your bike and doing some maintenance on your own, you'll get a sense of how advanced of a project you're ready to do.

Tool Time

When buying your first set of tools, don't succumb to the siren song of cheap, no-name tools. While a step above the pot metal atrocities usually included with your bike, cheap tools don't last long, and they damage parts much more frequently than quality ones. Also, tools do wear out and break, so having a lifetime warranty is a good idea. The top two brands of lifetime tools for a home user are Sears' Craftsman and The Home Depot's Husky. Both offer a "we-replace-it-if-it-ever-breaks" warranty. Although I have no experience with Husky tools, I can say that Sears has never rejected a broken tool I've returned, no matter how stupidly it was being abused when it failed. Yes, Snap-On offers a lifetime warranty, too, but you pay a premium for these tools. However, you always know where Sears and The Home Depot are, unlike the Snap-On delivery truck.

You can save a bundle by buying the complete tool sets, but you often end up with extras. Read the list of items carefully, since you don't need SAE tools to work on metric machinery. The absolute best time to buy tools are around "guy holidays" (no offense to female mechanics) such as Father's Day, the Fourth of July, and Christmas. You can save as much as 50 percent if you plan your purchases wisely. Also, Craftsman has a tool club that offers monthly discounts to feed your addiction.

While you can survive with a basic socket set, each of these items has a specific use. Note the colored tape used to mark the most commonly used sizes for easy identification.

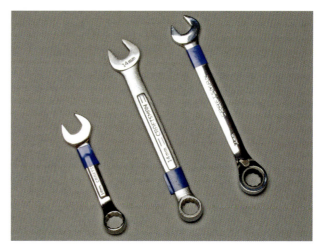

These wrenches are among my most commonly used tools. From right to left: Stubby combination, regular combination, and ratcheting combination wrench.

Basic Tools

Sockets and Drivers

At the very least you need a 3/8-inch ratchet and 8-mm, 10-mm, 12-mm, 14-mm, 17-mm, and 19-mm sockets. You'll also need a couple of socket extenders and a universal joint. These represent the absolute, bare minimum selection. Ideally, you should have 1/4- and 1/2-inch ratchets and sockets, too. The 1/4-inch ratchet needs correctly sized sockets ranging from 6 to 10 mm. A similar scope is recommended for 1/2-inch sockets, only at the big end of the spectrum from 14 to 19 mm and above. As you get further involved in wrenching, you'll find you need a set of deep sockets. Another nice addition is sockets with a universal joint built in. You'll be amazed at how much easier you can access exhaust manifold nuts and all sorts of fasteners tucked away in the bowels of your motorcycle.

A good rule of thumb when buying tools is that if you find yourself needing a single socket (or wrench) you should buy a set. Why? Simple arithmetic. When you buy the second one, you have usually paid the bulk of the cost for an entire set. So, if you find yourself needing that 17-mm deep socket, you can bet that before too long you'll want the 19 mm or the 14 mm. Go ahead, you know you want the set anyway.

Torque Wrenches

Another absolute must for a mechanic is a torque wrench—or two. Torque is a measurement of twisting force. You need a torque wrench because if you don't torque a fastener down tight enough, you risk having it vibrate loose. Go too far when tightening something, and you'll strip the threads or break the fastener.

Torque wrenches come in a couple varieties. The least expensive (and least useful) is a wrench that has two bars. The first bar is the hand grip and the second is the pointer. As you tighten a bolt, you bend the bar with the handle, moving a gauge under the pointer. Don't waste your money on this type.

You want to use a wrench that allows you to set the value you wish to torque a fastener. As you crank on the wrench, it will release with a click when you reach the specified torque. While you can buy torque wrenches with either foot-pounds or Newton-meters (SAE or metric), I recommend getting a torque wrench that has both measurement scales. Often you will get instructions with one torque figure but not the other. A dual-scale wrench saves you from having to convert the figures. If you can only afford one torque wrench, get one ranging from 20 to 100 or more foot-pounds. Buy one in inch-pounds (with a range up to 20 foot-pounds) as soon as you can.

Motorcycle fasteners are expensive, so don't count on "feel" when you're wrenching.

Even if you're using a torque wrench, you still have to use common sense when tightening nuts. For example, torquing down the bolts that secure a brake disc should be done in stages. Set your first torque value at about the halfway point, torque all the bolts in an alternating pattern, and then repeat at the full value. If you try to go for the full value at once, you can strip the bolts. Bigger nuts, like axles, don't need the intermediary step.

Also, unless otherwise stated, torque values are for dry threads. If you lubricate the threads, the lowered friction between the threads can allow you to exceed the recommended stretching force of the fastener, which can lead to damaged threads or broken fasteners.

One final word about torque wrenches: They are precise, expensive instruments. Don't use them as a breaker bar. It's tempting to turn the torque wrench all the way up and use the extra length to crack loose a stuck fastener. Unfortunately, you can ruin your torque wrench if you're not careful. Breaker bars have been designed solely for this purpose and are considerably cheaper to boot. Although they're usually as long as a torque wrench, you can lengthen them with a piece of pipe for those times you need even more leverage. Just be careful not to twist the head right off the bolt!

Wrenches

You can't work on a motorcycle without a set of combination wrenches. Again, you need them in the same sizes as the sockets mentioned above. Don't try to save money by buying wrenches that have two different-sized open ends on the same wrench. Combination wrenches have one open end and one closed end. The closed end has the same shape as a 12-point socket, making wrenching in tight places much easier than using the open end. As your collection expands, you'll find yourself lusting after a set of shorty combination wrenches. Give in to your desire. You'll be glad you did.

When you're really flush with cash, buy yourself a big honkin' adjustable wrench. Your buddies may tease you about needing a big tool (mine did). You'll have the last laugh, though. A big adjustable wrench can help you loosen any variety of axle nuts you may run into when no matching socket is handy.

Finally, stay away from the multi-tools you see advertised on TV. Most of them don't live up to their claims of fitting multiple sizes of nuts and bolts. Instead, they are equal-opportunity fastener manglers.

Allen Wrenches

While you can get by with a set of 3/8-inch Allen sockets, variety is a good thing to have in this class of tools. Personally, I prefer T-handle Allen keys for everything that doesn't require torquing.

However, you'll find the time when you need an Allen socket with a ball end. These allow you to operate the tool at a slight angle to the bolt head, allowing you to access some truly evil locations. Until recently, Snap-on was the only vendor I could find that made ball-end Allen sockets. However, Sears has just started selling a non-Craftsman branded set (meaning no lifetime warranty).

Since smaller Allen keys tend to wear out, you'll definitely want to make sure you have the warranty for your main set. Unfortunately, ball-end Allen keys tend to snap off if you get too aggressive with them. So, be careful and be prepared to buy more as they wear out.

Screwdrivers

This may sound crazy, but the shape of a screwdriver's grip plays an important role in how easy it is to use. Go to your local tool store and sample a few. Ones shaped with the natural curve of your palm with no edges to dig in to your hand offer the best grip.

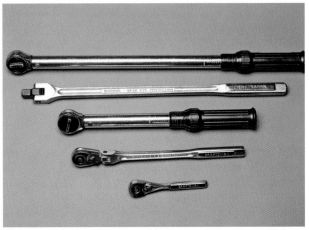

The more you get into wrenching, the more uses you'll find for different ratchets. A nice, basic selection consists of (from top to bottom) a torque wrench in foot-pounds, 1/2-inch breaker bar, torque wrench in inch-pounds, 3/8-inch flex ratchet, and 1/4-inch ratchet.

The Craftsman Professional series of screwdrivers are my favorite. The grips are composed of a grippy, slightly soft material that makes it possible to really crank on a stubborn screw. When the tips wear out, just take them back for replacements. Buy a set with at least three sizes of both flat heads and Phillips heads.

You'll also find jewelers' screwdrivers invaluable for working with tiny parts—even helping to pry out ornery circlips.

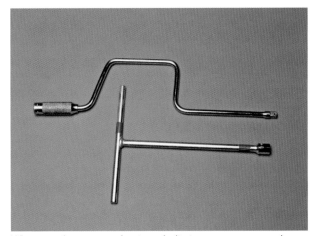

When you have several nuts or bolts to remove, a speed wrench or T-handle socket will speed things up tremendously. Which of these you choose depends on personal preference.

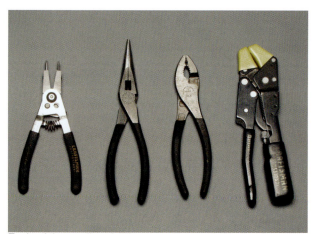

Each of these pliers fulfills a specific need: Circlip pliers, needle-nosed pliers, standard pliers, and locking pliers. Note the Soft Jaws slipped over the teeth of the locking pliers to prevent them from marring delicate surfaces.

Pliers

While you should use a socket or a wrench to tighten/loosen fasteners, pliers have a multitude of uses. Invest in a quality set. You need at least one standard, one needle-nosed, and one arc-joint pliers. Locking pliers (Vise-Grips) are versatile and handy and can perform a variety of jobs. I've even pressed a chain master link into place with a pair when I had no other tools available.

Hammers

Be wary of any time you're tempted to reach for a hammer when working on a motorcycle. A good rule of thumb is: If you are breaking a sweat or considering using brute force on a part, stop what you're doing, step back from the bike, and make sure you're not overlooking something. Always use the muscle between your ears before you start flexing the ones in your arms. You'd be surprised how many times you find some little thing that's keeping you from your goal.

Still, you will occasionally find that you need a hammer. A ball-peen hammer is ideal for rapping on an impact driver. You'll also find many uses for a rubber mallet or dead-blow hammer. Sometimes a little tap from one of these is just the trick for removing a stubborn side cover.

Electrical Tools

Troubleshooting electrical problems requires only a few tools. You need a good multimeter and continuity tester. In many cases, connectors are too far apart to use a continuity tester, and a homemade test light will do the trick. Also, you need a pair of wire cutters/strippers and a wire crimping tool—although, for truly permanent connections, you should solder the wires together. Electrical tape and zip ties will be called into action on more than just electrical jobs, so buy a bunch.

Drill

A hand drill, be it cordless or corded, plays a vital role in the home mechanic's toolkit. You'll also want to buy a variety of bits. Don't buy the super-cheap ones, either.

If you plan on using your drill for extracting stuck fasteners, you'll need something that generates some torque. I alternate between a light-duty battery-powered drill for little stuff and a beefy, corded model for major projects.

As with any tool that spins, wear safety glasses.

Impact Driver

Sometimes the only way to remove a stripped screw or stuck bolt is to give an impact driver a good whack with a ball-peen hammer. The force of the blow presses the driver into the fastener, thereby increasing the friction available to the bit that is being mechanically rotated by the impact.

This is a brutal, inelegant, but somehow satisfying tool when it works. However, when misused, an impact driver can do a remarkable amount of damage in a very short time. Wear eye protection.

Measuring Tools

Most basic maintenance work can be done with just a steel metric ruler, tape measure, and feeler gauges.

Once you move into more advanced projects, you'll find a dial or vernier caliper invaluable. If you're delving in to your engine's internals, you definitely want to buy the higher end models. While dial calipers are fairly easy to use, digital versions can often change from SAE to metric at the flip of a switch. Expect to spend some money on these. The same is true of dial indicators, which come in varying quality levels and measurement ranges. For most applications, a 1-inch range at 0.001-inch increments will suffice.

For a dial indicator to work, it needs to be solidly mounted to something. A magnetic base works great for this.

Cutting Tools

Sometimes you have to cut parts. If you don't have a rotary tool, a hack saw makes removing an endless drive chain much easier. Also, a matte knife and other assorted blades will find frequent use in your garage.

Files are also useful for deburring metal or plastic parts.

Toolbox

Big surprise, you need a place to keep your tools. Besides, it makes a great place to display all those stickers you get from the aftermarket companies when you buy their parts. Also, picking up and returning the tools scattered around and under the bike is a great way to regain focus when you're stymied.

While you can get by with a three-drawer toolbox with a basic tool set, you really need a way to organize your tools. Otherwise, you lose valuable wrenching time looking for a tool. Buy a free-standing rolling toolbox with a portable three- or four-drawer box on top of it. You can also buy a variety of shelves and other do-dads to increase your work space.

Little Necessary Extras
There are plenty of little things (some obvious, some not) that make working on your bike easier.

You'll need latex gloves. You're going to use some toxic chemicals, and your progeny will thank you for cleaning the parts, and not your hands, with contact cleaner.

Also, a good set of work gloves will serve two purposes. You'll keep your hands cleaner, giving you a better chance of ever laying hands on your significant other again, and like riding gloves, you'll have a better grip. While we're on the topic of hands, a tub of abrasive hand cleaner will make your transition from the garage to the dinner table a little easier.

Funnels of various sizes make fluid-related jobs easier. A big syringe lets you suck hydraulic fluid out of reservoirs or excess oil out of the crankcase if you overfill it.

A dedicated oil drain pan is a must. It should seal so that you can take the dirty oil to a recycling center.

A plastic dishwashing tub that will fit under your bike is nice for messy work. Similarly, small metal baking pans make sense for carburetors and other leaky parts while they sit on your workbench.

You never run out of uses for shop rags. While rags will come clean in a washing machine, unless you want to spend time scrubbing out the oily ring in the basin, you should just trash the dirtiest ones. Disposable shop towels work much better than regular paper towels. Also, during messy jobs such as changing oil put something under your bike to catch the splatter. Newspaper will work in a pinch, but little parts can roll under individual sheets only to be thrown away with the paper. Finish Line makes a great rubber-backed work mat to put under your bike. When it gets dirty, take it to your local pressure wash and give it a good spray.

One drawer in your toolbox will most likely begin to collect the smaller necessities. Mine has X-acto knives of varying styles, single-edged razor blades, Easy Out bolt extractors, and film canisters with miscellaneous nuts, bolts, washers, and cotter pins. I also have a collection of pens, pencils, Sharpies, and even grease pencils. Note pads are vital for keeping track of the order in which you removed parts. Paper tape (often found in photographic-supply stores) works great for labeling parts and wires without leaving a sticky residue when you remove it.

Also, a box of plastic sandwich bags will help you organize parts when you remove them. Take the time to label the bags. You'd be amazed what you can forget overnight.

Bike Supports
In a perfect world, we'd all have pneumatic bike lifts to raise our machines to a comfortable working height. The reality is that simple front and rear stands offer the stability for the

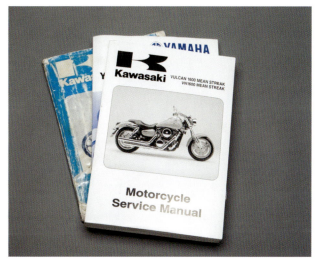

Factory service manuals are vital to the successful completion of any work on your bike. After awhile, they'll be dog-eared, grease-stained, familiar friends.

majority of motorcycle maintenance chores. Unfortunately, thanks to the shape of their swingarms, shaft-driven bikes don't like to rest on rear stands. Sears sells a hydraulic motorcycle jack that is ideal for cruisers, although you may need to fashion some chocks to keep the stand from pressing on vulnerable exhaust headers. Kawasaki sells a jack that can also support a variety of bikes via a selection of adapters.

Home mechanics have survived for years using a piece of wood and a scissor jack bought from a salvage yard. If you choose this method, make sure you have a backup system, like a tie-down thrown over a garage rafter, to keep gravity from winning.

As your skills advance, you'll find that you need to support your bike with the suspension off it. The Craftsman motorcycle jack (bottom) can lift your bike almost 2 feet off the ground. You may need to fabricate some wooden chocks to fit between the bike and jack. The Kawasaki bike jack (top) has a variety of adapters that mount on the top posts of the jack, enabling you to use it for a variety of purposes.

You can fill up several shelves in your garage with the consumables you use regularly when working on your bike. These are some of the most important ones.

You'll know you're really getting into wrenching when you decide you can't live without a set of air tools. Start with the basics: a compressor, a ratchet, an impact driver, a blower, and an air chuck.

BASICS

Nice-but-Not-Necessary Extras

Work Light
A variety of compact fluorescent units are available today. So, there's no reason you should ever have to use the annoying hot ones that rely on household light bulbs. If you don't burn yourself on the metal housing, you'll probably drop it and break the filament at an inopportune time.

Fluorescents are lightweight, cool, and tough.

Parts Washer
Many maintenance tasks require cleaning parts as an initial step. Hunching over the utility sink in your garage is tedious work. Consider buying a parts washer. While full-service shops have parts washers that sit on top of 50-gallon drums and pump cleaning fluid into a basin, you can buy a much less expensive home washer. Finish Line has a self-contained washer that holds a citrus-based solvent that you pump into a little tub with a hand-sprayer. When you're done cleaning, pull the plug and let the dirty solvent mix back with the rest inside the washer. When not in use, the washer can stand on end under your workbench.

Vise
While the need might not seem obvious at first, a bench-mounted vise plays a vital role in fork and shock work. Aluminum soft jaws are a nice addition.

T-handle Sockets or Speed Wrench
Removing a series of bolts goes much easier with a tool that allows you to spin it. Look in the pits at a national race, and you'll see T-handle sockets or speed wrenches. The T-handle allows you to spin a nut free by holding the center and spinning the T. A speed wrench operates on a similar means by giving you a rotating end and a handle in the middle to crank. Once you get used to either of these tools, ratchets seem painfully slow.

Tap and Die Set
Threads get stripped or damaged, but often you can clean them up with a tap or die without resorting to drilling them out. When all else fails, you can drill out the hole and tap it with the next larger sized threads. Applying tapping oil will make these tools last longer.

Screw Removal Kit
Fasteners break sooner or later, and when they do you'll have several options available to you. The ubiquitous Easy Out has been around for years, and when it works it can save your hide. Be forewarned though—when an Easy Out breaks, you'll usually need to take the damaged part to a machine shop.

Craftsman has recently introduced a line of Drill-Out extractors. These tools, on my initial experience, offer a solid alternative to Easy Outs. Craftsman's Bolt-Out removers fit over the heads of nuts and bolts and are also particularly effective.

Rotary Tool
From grinding off the rivets on a chain, to trimming your bodywork, to de-burring drilled parts, to cutting the heads off of stripped fasteners, a rotary tool and a variety of cutting wheels and grinding stones will see more use than you can imagine. Don't forget eye protection.

Air Compressor

At the very least, you want a refillable three- to five-gallon auxiliary air tank in your garage to top off your tires before a ride.

But forget about necessity: Have you ever seen how excited mechanics get over air tools? If you're one of them, you need about a 15-gallon capacity compressor and 150-psi maximum pressure for a basic garage setup.

The most important specification is the Standard Cubic Feet per Minute (SCFM) rating. Check to make sure that your compressor can handle your tools' requirements. However, you can get by on the low side of the requirements for intermittent use such as with a ratchet and impact driver. Painting, on the other hand, requires that your compressor is able to exceed the requirements. If you can't afford a complete air setup, cordless ratchets and corded electric impact drivers are available.

Specialty Tools

Depending on how far you get into this wrenching habit, you'll also find yourself collecting narrowly focused tools. A bead breaker and wheel-balancing tool falls into this category.

Specialty tools aren't all expensive, though. Fork oil-level tools, for example, come in a wide range of prices. Throughout this book, I've listed specialty tools required for a particular task in the "Tools" section at the beginning of each project.

How to Read This Book

While I'd be gratified if you read this book from cover to cover, we all know that our busy schedules may prevent that. So, the chapters cover groupings of projects based on a cruiser's different systems.

For the most part, the maintenance projects are listed at the beginning of each chapter. These projects are usually the easier ones and offer novice mechanics an opportunity to get comfortable wrenching.

To assist you in your planning, each project begins with an easy-to-read listing of information to give you an idea of what challenges you face. The items listed are:

Time

Many projects can be finished in a couple of hours. Others can take days. Wouldn't you like to know in advance? If you're a complete novice, you might want to factor in additional time. Old pros will probably breeze through some projects in significantly less time.

Tools

Although you can complete the majority of projects with the basic mechanics' tools listed in this chapter, each project requires a different subset of those tools listed here. Any special tools required are also noted. You should still check your factory service manual to make sure that your bike doesn't require a tool specific to it. Use this listing to justify your forays into your local tool store.

Experience

While bikes can vary from model to model, projects can be categorized into a broad range of difficulty. ★ means that almost anyone, no matter how little mechanical experience they have, should be able to complete the task with few problems. Each additional ★ marks an increased level of difficulty.

★ A project that a novice could complete.

★★ A project for a novice with a little help needed.

★★★ A project that requires a fair amount of mechanical expertise and comfort with complex assemblies.

★★★★ Only attempt this project if you are well versed in wrenching—perhaps if you've had some training. A project at this level is best attempted with the assistance of a more experienced mechanic.

★★★★★ A project best left to the pros, but those who aspire to professional tuner status could press ahead.

Cost

Each dollar sign indicates approximately $100. So, beginning with a single dollar sign, expect to pay up to $100 for the parts necessary to complete the project. If you see five dollar signs, the required components for the project will set you back $500 or more.

Parts

The listing of parts required to perform the modification is provided. However, you should still check your factory service manual to ascertain that your bike doesn't require anything special.

Tip

Knowledge comes from experience. The tip includes information to make the project easier, keep you from overlooking a little detail, or give you information about how the modification may affect your bike. You won't find this information in your factory service manual.

Benefit

This tells you what to expect from your time, money, and effort. It could be anything from a major boost in horsepower to a smoother running engine.

Complementary Modifications

This listing points to other projects that could help you get even more out of the current project. Again, you won't find this information in your factory service manual.

SECTION 2
GENERAL MAINTENANCE

Projects 1–6

PROJECT 1 | Adjusting Clutch and Throttle Freeplay

by Eric Goor

Time: 30 minutes

Tools: Phillips screwdriver, open-end wrenches, and (maybe) needle-nosed pliers

Talent: ★

Tab: $0

Parts: None

Tip: If you run out of adjustment range up at the lever, try adjusting the cables down by the engine.

Benefit: Precise throttle control gives you maximum flexibility in the on/off/on throttle scenarios you encounter when entering a corner, or riding a series of them.

One hallmark of a skilled rider is the ability to precisely deliver the right amount of throttle at the right time. Smooth transitions on and off the throttle play a vital role in keeping the chassis stable in a corner, and a stable chassis will enable you to maximize your cruiser's ground clearance. Since the throttle cables are the direct link between your hand and the butterfly valves in the mixers, minimizing slop in the system will pay big dividends. Riders of large-displacement V-twin cruisers will especially benefit, as their compression braking can be quite dramatic. The same level of control is desirable for bikes with cable-actuated clutches. Whether you're trying for a smooth launch from a stoplight or flawless downshifts with a passenger on the back, you want your inputs to be seamless.

Throttle Freeplay
To check the throttle freeplay, hold the grip between your fingers and roll it back and forth until you begin to feel the pull of the cable. Pick a spot on the grip and watch it to measure the freeplay. If you have trouble visualizing the measurement, hold a metric tape measure up to the grip.

Most factory service manuals will tell you that 2 to 3 mm is the correct amount of throttle freeplay.

If you've determined that the freeplay needs adjustment, loosen the locking nut(s) near the throttle grip. Some bikes will only have one adjuster. For two-adjuster models, loosen the nuts until there is plenty of slack in the system. Next, tighten the deceleration adjuster (the cable that pulls the grip into the throttle-closed position) so that there is no slack when the throttle is held closed. Tighten the deceleration locking nut. Now, adjust the acceleration cable's adjuster until the desired amount of freeplay is present in the grip, and tighten its locking nut. Ensure that there are plenty of threads (at least three) engaged in the adjuster body.

If you can't get the proper amount of freeplay with the adjuster(s), set the adjuster(s) to the middle of its/their range and adjust the cable down by the carburetors/throttle bodies. Begin by removing the tank and any bodywork that will interfere with your access to cables. On some bikes you may need to remove or disassemble the air box to reach the bell crank. Locate the adjuster nuts for the throttle cables where they attach to the throttle body. Loosen the lock nut on the

Rock the grip back and forth to measure the throttle freeplay. Use a metric tape measure for an accurate reading.

deceleration cable and adjust the cable until there is no slack with the grip in the closed position. Now, adjust the freeplay of the acceleration cable to spec in the same manner. Any final fine-tuning to get the freeplay to your personal preferences can be done at the throttle grip end. Adjusting the throttle freeplay using this method is time consuming, but pays off when you need to correct the freeplay in the future. When you are satisfied with the cable settings, tighten the lock nuts firmly to prevent them from vibrating loose.

Food for thought: While 2- to 3-mm freeplay may be the factory spec, many riders prefer even less freeplay, giving them the feeling of a seamless connection to the carburetors or injector housings. Experiment with different freeplay amounts to find the setting that suits your riding style. A word of warning about using less freeplay than the factory specifies: If the throttle cables are too tight, they can cause the throttle to stick, close very slowly, or not close complete-

ly, so check thoroughly by rolling the throttle open and releasing it from a variety of settings. Finally, run the engine at idle speed and turn the handlebar to both the right and left to make sure that the engine speed does not change. If it does, check the cable routing and freeplay again.

Clutch Freeplay

For cruisers with hydraulic clutches, you can skip this section because hydraulic systems are self-adjusting. Cruisers with cable-actuated clutches should be checked regularly, though. Also, the clutch-lever freeplay adjustment can accommodate various rider preferences and hand sizes. (If you have hand size problems with a hydraulic clutch, try buying an adjustable lever.) To measure the freeplay, pull in the clutch lever to take up the slack in the cable. Now measure the gap between the clutch lever holder and the lever itself. Again, most manufacturers recommend 2 to 3 mm of freeplay.

Set the freeplay with the cable adjusters on the throttle cables. If the bike has two adjusters, set the deceleration cable first.

To adjust the freeplay, loosen the knurled lock screw on the clutch-lever holder. (If your bike doesn't have one, look for an inline adjuster somewhere in the middle of the cable.) Now, unscrew the adjuster for less slack or screw it in for more slack. Riders with smaller hands will probably want to have a bit more slack than those with larger hands. Also, depending on where the clutch engages in the lever travel, you may want to adjust it to engage at a different point. If you give the lever extra freeplay, make sure that the clutch releases fully when the lever is pulled all the way in. If it doesn't, your ability to shift smoothly will be compromised, and the transmission will undergo unnecessary stress when you downshift. If the freeplay is less than the recommended amount, the clutch may not fully engage, causing clutch slip and premature clutch wear.

Sometimes cable stretch makes it impossible for you to get the proper clutch freeplay. If this happens, turn the adjuster on the lever holder so that 5 to 6 mm of the thread is visible. Next, adjust the slack at the lower end of the cable. Slide the cable dust cover out of the way, if there is one. Loosen the nuts as far as they will go. Now, pull the cable tight by sliding it inside the bracket. Tighten the nuts firmly enough so they will not vibrate loose, and return the

Space is usually pretty tight, so take your time as you adjust the cables on the bell crank.

dust cover to its proper position. The freeplay can now be adjusted by the screw at the lever. You've just officially outsmarted your motorcycle.

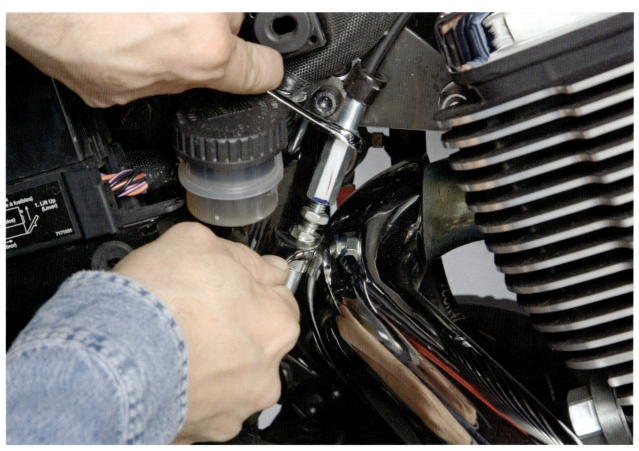

Bikes that don't have an adjuster up by the lever will have an inline one that looks something like this.

Take up the slack in the clutch cable and measure the gap between the lever holder and the lever.

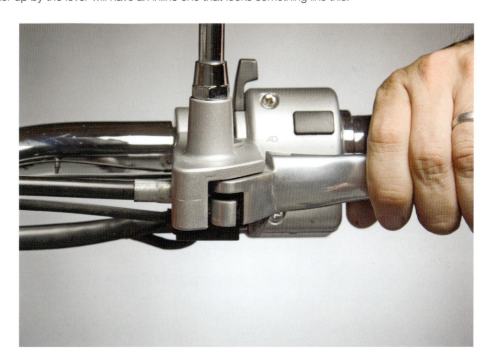

GENERAL MAINTENANCE

17

PROJECT 2 | General Lubrication

Time: 30 minutes

Tools: Assorted wrenches, cable lubrication tool, cable lube, chain cleaner, molybdenum-based grease, WD-40, small paint brush

Talent: ★

Tab: $

Parts: None

Tip: Apply a thin coating of grease to cables where they attach to the controls.

Benefit: Silky-smooth control operation.

In most owner's manuals, the maintenance chart lists general lubrication as one of the activities. Since few people actually read their owner's manual, and even fewer follow the periodic maintenance intervals, some important motorcycle parts will degrade quicker than they should due to simple neglect. For the average rider, lubing your bike three times a season (the beginning, middle, and end) will suffice. If you ride the wheels off your motorcycle, give it a quick lube job every 7,500 miles. By performing this simple maintenance on a regular basis, you prolong the life of the bike's components, ensure that everything works the way it should, and maybe even catch a problem as it starts.

Lubing your motorcycle's cables should take no more than 15 minutes and pays dividends every time you operate a control. For the throttle cables, unscrew the throttle housing on the grip and adjust the cables for maximum slack. After you release one of the cables, the other will slip right off. For the clutch cable, screw the adjuster all the way in for maximum slack but line up the slot in the adjuster with the slot of the lever holder. You should be able to pull the end of the cable free of the adjuster and release the cable. Some bikes require that the clutch lever be removed from the mount before you can free the clutch cable.

For quick work on the cables, nothing beats a pressure cable luber from accessory companies such as Motion Pro or Lockhart Phillips. Basically, you clamp a rubber stopper over one end of the cable, insert a tube from a can of silicon-based cable lubricant into a little hole, and give the nozzle a

The cable luber forces the lubricant through the cable, making sure the entire length of the cable is protected. Do not use chain lube on cables.

Apply a protective coat of grease to all exposed sections of cables. Don't forget to lube the fittings so they will move freely within their mounts.

Before lubing pivot points, wipe them clean of any dirt or grit. Keep the threads clean and dry. Retorque the fasteners to the proper spec to keep them from vibrating loose.

squeeze. The can's pressure forces the lubricant through the cable. Apply the lubricant in short bursts until the bottom end of the cable begins to bubble or drip. A well-placed rag can catch the drips before they make a mess of the engine.

Before you reassemble each cable, be sure to apply a dab of grease to all the places the cable might rub. The fittings at the cable ends need grease, and any exposed sections of the cable should receive a protective coating, too. A small paint brush will help you grease parts in tight places.

Don't forget the choke cable and the speedometer cable. Lubricate the choke cable like all the others. Mechanical speedometer cables (a dying breed but still around on many bikes) should be unhooked from the speedometer. Pull out the inner cable and pack the cable top with molybdenum grease. Slide the cable back into place. Any time the front wheel is removed, pack some moly grease into the drive mechanism.

Now go over your bike and lubricate every part that moves. Unscrew the handlebar lever pivots and brush on some grease. Apply a couple drops of oil to the side stand pivot. Give the rider and passenger pegs a quick squirt of WD-40 or oil. Don't forget to spritz floorboard mounts. Remove the shifter pivot and clean any grit out of the works. Apply grease to the pivot, but be sure to keep the pivot's threads clean and dry. Some manufacturers recommend a drop of non-permanent thread-locking agent such as Loctite on the threads to make sure the pivot bolt doesn't back out after reassembly. Be sure to torque it to the proper spec. Follow the same precautions for the brake pedal.

While most cruisers have shaft or belt drives, you still find chains on many smaller displacement models. Save lubing the chain until the end of a ride when the chain is hot—the lube soaks into the rollers better. Before you lube the chain, spray a clean rag with WD-40 and wipe all the dirt and grit from the chain. If your chain is really grungy, Motorex makes a chain cleaner that will strip the gunk without harming the O-rings. It even smells nice. Do not succumb to the temptation of using the running engine to rotate the chain while you wipe it. Many mechanics with missing fingers (or parts of fingers) can attest to the foolishness of this technique.

GENERAL MAINTENANCE

All moving parts such as peg pivots and side stands will benefit from a shot of WD-40 or a few drops of oil to keep them working freely. If you're particularly fastidious, you can disassemble the parts and grease them.

Once you've cleaned the chain, apply a coat of quality chain lube to the space between the links where the O-rings reside, spraying from the inside run of the chain to allow centrifugal force to push it through to the other side. (Motorex makes small, handy chain-lube containers that make it easy to get the spray where you want it. The little containers are then refilled with a large can—very clever.) Don't worry if you spray on too much chain lube. After the lube has had a chance to set, but before you ride your bike again, take another clean rag dampened with WD-40 and wipe off the excess lube before it has a chance to get flung onto your shiny chrome wheels. Lubricate your chain every 400 miles or so. Any time your bike is ridden in the rain or exposed to salt spray from the ocean, you should lube the chain as soon as possible.

Lubricate the chain when it is warm, but then let it sit until it cools off and the lube sets. To help keep your wheels clean, wipe off the excess lube before your next ride.

PROJECT 3

Oil Change

Time: 1 hour

Tools: Oil filter wrench, oil catch pan, wrench for drain plug, contact cleaner, rags for cleanup

Talent: ★

Tab: $

Parts: Oil filer, drain plug gasket, several quarts oil (check your factory service manual)

Tip: Be earth friendly. Find your local oil recycling center at www.earth911.org

Benefit: You may not feel it, but your engine will—fresh oil means less friction means better engine operation and power.

The topic of motor oil for motorcycles can stir up some heated arguments. Some riders insist that the pricier, motorcycle-specific oil is nothing more than a way to extort money out of the riding public. These nonbelievers claim that three generations of their riding family (or a riding buddy's family) have used whatever lubricant was on sale down at the local Econopart without any mechanical problems. Well, oils have changed more in the last 10 years than they have in the previous 50, so what may have worked before doesn't necessarily apply to the current, highly tuned motorcycles. Today, even air-cooled V-twins are machined to tolerances much higher than previous

The limited space on some cruisers often leaves little room for removing a filter by hand.

You can't remove a drain plug without getting oil on your hand. Be smart and wear a latex glove, which is cheap, disposable, and doesn't leave your hand filthy and raw.

generations of engines. Just because the engine looks retro doesn't mean it isn't high-tech. Use motorcycle oil.

Why don't we want to use automotive oil in motorcycles? First, and most importantly, motorcycle engines share the oil with their transmissions and the clutch. The shearing forces applied by the transmission and clutch will wear out cheap automobile oil much faster than motorcycle oil. Also, the newer "low friction" or "energy conserving" automobile oils use friction modifiers that may have a negative effect on your clutch's ability to engage properly. Several motorcycle manufacturers have even gone so far as to recommend that you not use the SJ-labeled auto oils. Additionally, motorcycle oils usually have higher concentrations of beneficial detergents and other compounds that are not present in automotive blends since they would damage catalytic converters. (As catalysts become more common on bikes, this feature of motorcycle oil may change.) In short, motorcycle oils are designed for the specific conditions that bike engines exert on them.

The next choice you will face is whether to run petroleum-based or synthetic oils in your engine. Although the lines between the two categories have blurred as dinosaur juice, man-made oils, and additives have all been blended into different types of products, synthetic oils offer some clear advantages. First, synthetics offer more consistent viscosity across a range of temperatures. They thin out less when hot and thicken less when they cool. Second, synthetics transfer heat better than petroleum-based lubricants, so your engine should run cooler. Finally, synthetic oils offer lower volatility and better oxidation stability. This means if your bike overheats for some reason, a synthetic oil stands a better chance of not cooking away, leaving the metal of engine parts vulnerable.

When it comes to actually changing the oil, two schools of thought reign: The first, and more common, states that the engine should be warmed up to operating temperature prior to draining. Mixing up all the crud that would settle out of the oil mixture when the bike cools, helps flush it out more easily. The other camp stresses that if the engine hasn't been operated in more than 24 hours, the contaminants are waiting in the oil pan. Why not just drain the gunk without redistributing it throughout the labyrinth of oil passages? Regardless of which route you follow, make sure you can loosen the oil filter before you heat up the engine and/or drain the oil. You don't want to burn your knuckles on the header struggling to turn the filter or be forced to run that nice, clean oil through a dirty filter. If you can't loosen it by

Lube the filter's O-ring with fresh oil to help it get a good seal with the engine. Make sure the contact surface is clean, too.

hand, you need to invest in a filter wrench. Channel locks can work, in a pinch.

Novice mechanics shouldn't worry about violating a new bike's warranty. Just save your receipts and keep a record of the date and mileage of each change. Take your time and follow these steps.

Park your bike on a level surface. Putting your bike in gear to keep it from rolling off the stand while you work on the drain plug would be a smart move. (Note: Some manufacturers will recommend that your bike be straight up and down and not on its side stand. A jack will help you hold it in position.) Wear latex gloves during the messy part of the oil change. Used motor oil is a known carcinogen, and despite claims to the contrary, women do not dig greasy fingernails. Locate the drain plug on the bottom of the oil pan and carefully loosen it. If the plug gives you trouble, brace the bike so it doesn't roll off the stand before you give the wrench a yank or resort to a breaker bar. Drain the oil into a container suitable to transport to your local oil recycling center. Pouring old oil into the ground is irresponsible, illegal, and just plain bad juju and will force your descendants to drink funny-tasting water. Unless your catch pan is large enough to catch both the oil from the drain and the filter at the same time, wait until the bike's oil pan is empty before removing the filter. Remove the oil filter with a filter wrench (or by hand if it's loose enough). If your metal band-style wrench is too large to grip the small diameter of your bike's spin-on filter, try the old trick of folding up a rag under the wrench. Don't forget to pour out the remaining oil from the filter into the catch pan. If your engine has an oil screen separate from the oil filter, you may need to clean it in solvent. Check your factory service manual to make sure.

When you choose your oil filter, you can't go wrong with the original equipment replacement. However, if you stick to the major brand filters, you should be OK. You can even buy chrome oil filters for cruisers that have them mounted in visible locations. If you find a filter that is significantly cheaper than every other filter you've seen, be careful. Cheaper isn't always better when you consider how vital clean oil is to your engine.

Using your finger, wipe a film of fresh oil on the filter's O-ring. Prior to screwing the filter into place, wipe down the gasket's contact surface on the engine and make sure there is no grit anywhere that might break the O-ring's seal. Follow

Oil pans are expensive and difficult to replace or rethread. Use a torque wrench to make the drain plug stay put without risking stripped threads.

the filter manufacturer's specifications for tightening the filter. Use a new drain plug washer if necessary, and torque it to the factory specification. Don't rely on your torque-elbow and risk the embarrassment—and expense—of stripped threads. Fill the engine with the amount and type of oil recommended in the owner's manual. Before starting the engine, wipe down all the engine's oily surfaces. If you're in a well-vented area, give it a quick squirt of contact cleaner. This way oil leaks will be easier to spot after the engine has been run for a short while.

Now it's time to take the bike back out of gear and start it up. When you first start your engine, don't be alarmed if the oil light stays on a little longer than usual. The filter needs to fill with oil. Some fussy riders crank the starter and kill the engine when it fires, repeating the process until the oil light shuts off, in an effort to avoid running the engine while dry. Others crank the engine with the kill switch off until the light disappears. Once the engine reaches operating temperature, shut it down and wait a minute or so before checking the oil level. You may need to add or subtract oil as necessary, since the filter holds a few ounces of oil. Check for leaks, and you're done. Now, go for a ride.

If your bike has an oil screen, don't forget to check the screen for metal shavings that might signal engine problems before you clean it with a high-flash-point solvent.

Shaft Drive Housing Oil Check/Change

Unlike chain or belt drives that require occasional adjustment, shaft drives just need their lubricant changed as part of your bike's long-term maintenance. For example, after the initial 600-mile fluid change, the periodic maintenance schedule for the Mean Streak shown here recommends the next change be at 24,000 miles. That doesn't mean you should forget the shaft drive until the next change. You should still inspect the housing for leaks and check the oil level at the bare minimum of every other engine oil change.

Checking the gear case oil level is pretty simple. Make sure your bike is straight up and down. Using a really big flathead screwdriver, remove the oil filler cap. The oil level should be at the bottom of the opening. If not, double check for leaks before topping off the hypoid-gear oil with the manufacturer's recommended viscosity.

If you're going to change the oil, take your bike for a ride to heat up the oil. Once your bike is level on a stand, remove the filler cap and drain plug. (You can store the used oil in the same container as your engine oil.) Clean up any oil spills that could end up on your tire to prevent Bad Things from happening. Install a fresh drain plug gasket and torque the plug to spec. Fill the gear case to the bottom of the filler with the manufacturer's recommended viscosity oil. Close the filler. After your next ride, check for leaks. Pretty simple, eh?

This gear case is ready to go until the next inspection.

PROJECT 4

Coolant Check and Change

Time: 15 minutes to 1 hour

Tools: Sockets, ratchet, wrenches, screwdrivers, funnel, rags, antifreeze tester, drain pan

Talent: ★

Tab: $

Parts: Coolant, new copper drain-plug washer, Water Wetter (optional)

Tip: When adding water to the cooling system, only use distilled water.

Benefit: With Water Wetter, your engine should run cooler.

Coolant plays a vital role in the health of your engine. Unfortunately, when many riders think of coolant, they only think of "antifreeze." While protecting your engine from freezing during winter storage is important, antifreeze (coolant) also performs several other vital duties. The aluminum internals to your engine are prone to oxidizing. Coolant and other products, such as Water Wetter, form a protective coating over the bare aluminum, keeping it from eroding at high-heat areas

Check your antifreeze every fall, particularly if your bike is stored in an unheated garage. These test strips don't take up any room in your toolbox.

and from building up on cooler locations, which would reduce the efficiency of the cooling system. Coolant also lubricates the water pump and prevents foaming.

Every fall, before the temperatures begin to drop, you should test your antifreeze to make sure it will handle the cold. Changing the coolant every two years is also a good idea. Although the overflow tank is the easiest to access for your annual coolant test, why not spend the extra time to expose the cooling system filler cap so you can test what's actually inside the engine? So, with your engine cold (a hot cooling system is under pressure and will spew potentially scalding coolant all over the place), remove the filler cap. For the actual testing, you have a couple of choices. Prestone makes clever throwaway test strips that you dip into the coolant, read the color, then toss into the trash. You can also use a testing tool that measures the specific gravity of the solution and displays the results with floating balls or a needle. While you're at it, check the color of the coolant. If it looks like green like Mountain Dew, it's probably OK. Other colors may signal engine problems such as rust residue (red-brown) or oil residue (black).

So, what if your coolant didn't pass muster? Perhaps you added straight water to your overflow tank when you noticed it was low, or maybe the coolant is just old. Why gamble? Since it's so easy to replace the coolant, just do it. Then you won't have to wonder if you added enough antifreeze to bring the protection up to the proper level.

Begin with your bike in gear on its side stand. Locate the drain plug (it's usually on the water pump cover). Place a container large enough to hold all the coolant under the plug. If you want to keep the antifreeze off your skin, wear latex gloves. Open the filler cap at the top of the system. Using a wrench, unscrew the plug. Before the plug is completely out, the coolant will start to leak past the threads, so be prepared. Pull the plug free and let the system drain. As soon as the system is completely empty, reinstall the plug and a fresh copper washer and torque to specs. Next, empty the expansion tank into the catch pan. Pour the used antifreeze into a suitable container for transporting it to a recycling center or auto parts store. For some reason, children and pets have a fatal attraction for this extremely toxic liquid, so get it sealed up and out of reach quickly.

Those who will be filling their cooling systems with a 50/50 mix of coolant and distilled water should mix the solution prior to pouring it into the filler. That way you're certain about the mixture no matter how much liquid the system requires. Also, you'll have a container of the proper mixture if you need to top off the expansion tank in the future. If you want your cooling system to be more efficient, borrow a trick from the road race guys and buy a bottle of Red Line Water Wetter. You should notice a slight drop in operating temperature with Water Wetter in your coolant. Also, if you truly care about the condition of your engine,

Coolant usually pours out of the drain plug with a good deal of force, so be prepared. Make sure you have a container that will hold all your bike's old coolant, or you'll find out how hard it is to clean up antifreeze. Immediately transfer the poisonous liquid to a sealed container.

only use distilled water. Tap water will have varying levels of minerals (depending on the community) that can create deposits on the engine internals.

Once you've filled the cooling system to the brim, you need to run the engine with the filler cap off. As the engine warms up, you'll see bubbles working their way out of the system. In fact, as the engine circulates the coolant, you may see the level drop quite a ways. Keep topping off the system as the level drops. When the engine starts to warm up, the coolant will start to expand out of the filler. Stop the engine and replace the radiator cap. Fill the expansion tank until the level is midway between the two lines. Now, take your bike for a short ride to get it completely up to temperature, then park the bike and allow it to cool off completely. Top off the cooling system and button up your bodywork. You should now be good to go for at least another year.

PROJECT 5

Cleaning/Replacing Air Filter

Time: 30 minutes or a couple hours if air-drying a filter

Tools: Wrenches, sockets, ratchet, #2 Phillips screwdriver, rear stand, clean rags, compressed air

Talent: ★

Tab: $

Parts: OE or aftermarket air filter, air filter cleaner, air filter oil

Tip: Buying a washable aftermarket filter will pay for itself in a couple of cleanings.

Benefit: Consisten fuel mixture, better airflow

Complementary Modifications: Rejet carbs (see Project 23), adjust EFI (see Project 25), synchronize carbs/throttle bodies (see Project 22)

Not too long ago, a great way to increase your bike's horsepower was to toss the stock air box, install a set of pod filters, and rejet your carbs. Well, bike manufacturers have gotten better at letting the engines breathe while still keeping intake honk to a reasonable level. One thing hasn't changed, though. For your engine to operate at peak efficiencies, the air filter needs to be clean to allow maximum airflow. Let your filter get dirty, and you'll experience power loss, reduced gas mileage, and possible plug fouling. Regular—or at least annual—cleaning of your bike's air filter is a simple way to keep it running great.

Gaining access to your air filter depends on where the manufacturer has hidden it. The job may be as simple as removing the cover on the side of the engine, or you may have to remove the tank to get to the filter. Consult your owner's or factory service manual to see if the seat or any bodywork needs to be removed prior to unbolting the tank. If your bike requires that the tank be removed, be sure to turn the petcock to the off position and place a rag below the fuel line you'll be disconnecting. Remove the bolts securing the tank and make sure you disconnect all the hoses and wires. Label any hoses that aren't already color coded to ease reassembly. Some bikes

Although this is not as critical with a V-twin with the filter hanging off the side as with a V-four with a more traditional air box, you should keep the dirty side of the filter away from the throttle intakes.

Rinsing from the clean side out, use cold water to flush out the dirt, oil, and solvent. Let the filter air dry. Don't use compressed air or a hair dryer.

Oil the filter one pleat at a time. Let it sit for 20 minutes, then re-oil any white spots in the cotton.

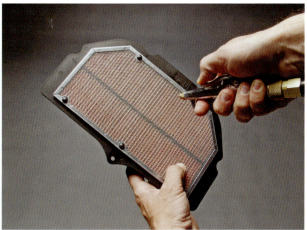

With paper filters, you should tap the filter on a table or trash can to knock the big chunks of crud free. Remove the rest of the dirt by blowing from the back side of the filter.

with tank-top instrumentation require that the housing be removed before lifting off the tank, while others can be left in place. Check that factory service manual!

Lift up the tank and place it securely out of your way. Setting it on an old tire works well and prevents nicking the paint as you set the tank on the garage floor. Give the top of the air box a quick blast with compressed air to clear out any dirt that may have accumulated in the screw holes. Remove the Phillips-head screws on the air box cover, making sure that you don't drop any into the depths of the engine compartment, never to be seen again. (If this happens to you, sometimes the part can be retrieved with a magnetic pickup tool—if you're lucky.) Once all the screws are removed, remove the lid and inspect the air box to determine the path of airflow in from the outside. Carefully, remove the air filter, making sure you don't knock any grit into the clean side of the air box. If the top of the air box is going to be off for more than a minute or so, cover the throttle intakes with clean rags or paper towels.

The kinds of air filters used vary by manufacturer. Generally, you will clean foam filters in a solvent. Paper filters can be blown out with compressed air from the back side of the filter. The best idea, though, is to buy an aftermarket filter such as BMC or K&N. These pleated cotton filters are reusable and should last the life of your motorcycle. The filters are constructed out of cotton that is trapped between two sheets of wire mesh. The manufacturers make sure that the filters will not require rejetting. So, they are true drop-in replacements to the OE versions. While they may cost more to replace, you will save money over the years that you own your bike.

Cleaning a reusable filter is a four-part operation. First, spray or pour on the solvent to cut the oil that is used to trap the dirt and let it soak for a few minutes. Rinse out the oil from the back side (the clean side) of the filter with cold water until the cotton fibers are clean. Dry the filter by placing it in the sun for a few hours or hanging it in your garage overnight. Do not use compressed air or a hair dryer or you will shrink the cotton and render the filter useless. When it's dry, you want to coat the cotton with filter oil. (Never use any other kind of oil but filter oil.) If you are using a spray, one pass per pleat will suffice. For squeeze bottles, make one pass per pleat in the bottom of the pleat. Don't saturate the filter. Let it sit for 20 minutes and re-oil any white spots on the cotton.

Installing a fresh or cleaned filter is as easy as its removal. However, you should make sure that you have the filter facing the correct direction. For example, paper filters will generally have a screen to support the back of the filter to keep it from flexing or tearing during operation. If your air box has an O-ring to seal the filter access, ascertain that it is in position. After you close the air box, take a moment to make sure that it sealed properly. Screw the cover in place in an alternating pattern (rather than going around the circumference) to make sure that the pressure is evenly applied. Reinstall the tank, bodywork, and seat as necessary—and breathe easy.

PROJECT 6

Spark Plug Check and Replacement

Time: 1 hour

Tools: Sockets, screwdrivers, plug wrench, wire gap gauge, rags anti-seize compound, compressed air, brass brush, magnet to retrieve plug

Talent: ★

Tab: $

Parts: Spark Plugs

Tip: Blow off the top of the engine with compressed air to keep abrasives out of the cylinders.

Benefit: Smoother-running engine

Spark plugs live a hard life: high pressure, extreme temperatures, and constant explosions. Should we be surprised that the OEs expect us to check the conditions of our bike's plugs at every service? If you look at your owner's manual, you'll find that should be just about every 4,000 to 7,500 miles. Fortunately, the plugs don't often need replacing, just a quick brush off and back into the chamber of horrors they go. Still, you shouldn't let spark plugs' hardiness lull you into a false sense of security. Few things can wreak havoc on your engine's performance like a fouled plug.

Although you often read about race tuners reading plugs on bikes fresh off the track, we're not trying to squeeze that last 1/10 horsepower that could mean the difference between standing on top of the podium or somewhere else. Right now we're concerned with making sure your bike is running as it's designed to—the performance modifications come later.

Your bike's engine should be cold when you check the plugs. Place your bike on its side stand and remove any necessary bodywork, the seat, tank, or air box required to gain access to the plugs. Some bikes (particularly those V-fours) may require a few tricks to gain access to the plugs, so check your factory service manual first. While the air box is off, the carburetors are vulnerable to dirt, so cover them with clean rags or paper towels. Once you have access to the plugs, blow the top of the head off with compressed air. Now, remove the plug wire from one plug at a time. Carefully label each wire with a piece of tape. Although it sounds obvious, you don't want to risk mixing up the plugs and their wiring—Bad Things could happen.

Don't just jump in and remove the plug once it's exposed. Even though a cap has been covering the spark plug well, you should give it a quick blast with compressed air, too. You'd be surprised how frequently sand or pebbles pop out, and you sure don't want them in your engine. Once you've given it a shot of clean air, now you can safely remove the plug. The tight quarters around the engine's head may require some patience to remove the plug. If you've unscrewed the plug completely and can't get a grip on it to pull it out, try using a magnet to retrieve the wayward part.

Look closely at the electrode and insulator. It should be a light tan or gray color. Check the sidebar "Reading Your Plugs" for tips on what other colors mean. Using a brass brush, clean the plug of any deposits. Next, measure the gap with a wire thickness gauge. The flat spade gauges don't give accurate readings unless held perfectly square to

Despite the manufacturer's best efforts, sand and pebbles work their way into remarkable places. Blow off the cylinder head before removing the plug caps, and blow out the plug wells, too.

Reading Your Plugs

So, you've taken a plug out of its cubby hole, and you're wondering what to do with it. Impress your friends and neighbors by emulating famous race tuners: Closely examine the plug to see what it tells you about the state of your engine. Of course, you'll need some sort of comparison. The photos below will provide examples of what to look for. Remember, these are, for the most part, extreme examples meant to clearly illustrate the various plug conditions. Your plugs, unless they have a perfect tan, will probably have subtler symptoms.

You want all your plugs to look as pretty as this one. Note the nice, even grey or tan on the insulator. The electrode exhibits only slight erosion.

Carbon fouling shows up as dry, soft, black soot on the insulator and electrode. Although usually caused by a too-rich fuel mixture, other potential problems are shorting ignition leads and too-cold plug temperature. A badly carbon-fouled plug can lead to difficult starting, misfiring, and uneven acceleration.

This plug has been subjected to severe overheating as illustrated by the extremely white insulator and small black specks. If you look closely, you will see more electrode erosion than on a normal plug. When a plug overheats, the engine experiences a loss of power under heavy loads, such as at high-speed, high-rpm running. An overly advanced ignition, too-hot plug temperature, or poor engine cooling could be the culprit. Improperly torqued plugs can also overheat.

Oil fouling is characterized by black, gooey deposits on the insulator and electrode. Have you noticed that your bike was difficult to start and that the engine missed frequently at speed? The oil came from somewhere, and the likely culprits are worn piston rings or valve guides.

Photos courtesy of Denso Sales California, Inc.

If you must remove two or more plug wires at once, wrap them with tape and number them unless the factory was kind enough to do it for you.

the gap. (Don't even bother taking the round ramped gauges out of the toolbox. In fact—throw them away. They're worthless.) If the gap is too narrow or too wide, use the gapping tool usually attached to a set of wire gauges to carefully bend the side electrode outward. When you're using force to bend the electrode, a little effort goes a long way.

When the gap is correct, wipe the threads on a clean rag, then apply a little anti-seize to them. Carefully, insert the plug into the plug hole. If you can get your fingers down into the plug well, rotate the plug counterclockwise until you feel the threads drop into synch. Then, using only your fingers, rotate it clockwise to engage the threads. You can do this with the plug socket, but your feel is much more limited. Cross-threading a spark plug (or any fastener, for that matter) can be a time-consuming and expensive mistake. Screw in the plug finger-tight, and then snug it down with a torque wrench. This step is vitally important, since some plug failures are associated with incorrectly torqued plugs.

Before reassembling your entire bike, start the engine to make sure all the cylinders are firing correctly. (Fuel-injected bikes may need the gas tank installed to maintain proper fuel pressure.) If anything sounds amiss, check all plug connections and also make sure all the wires are connected to the correct cylinders. When everything sounds right, button up your bike and go ride another 4,000 miles.

Use a brass brush to clean the electrode and insulator. Be sure to clean the plug again with compressed air before you reinstall it. Note: Some fine-tipped plugs cannot withstand this treatment.

The best way to check a plug's gap is with a wire tool. This eliminates the requirement that the tool be held perpendicular to the electrode.

Spark plugs can sometime be ornery when you try to remove them because of cold welding between the aluminum of the head and the steel of the plug. Anti-seize helps to remedy the problem.

GENERAL MAINTENANCE

SECTION 3
BRAKES

Projects 7-13

PROJECT 7 — Brake Pad Change

by Eric Goor

Time: 1 hour

Tools: Wrenches; sockets; flathead screwdriver; Allen keys; torque wrench; rags; 60- or 80-grit sandpaper; toothbrush or other small cleaning brush, a thin piece of wood; brake/contact cleaner; organic cleaner such as Simple Green; waterproof, high-temperature grease; bike lift (optional)

Talent: ★

Tab: $

Parts: Brake pads

Tip: As long as you have calipers off, clean the pistons to spot problems early and ensure proper brake operation.

Benefit: Better stopping power

Complementary Modifications: Change brake fluid (see Project 9)

Brakes perform the most important job on a motorcycle. Consequently, you should pay special attention to the condition of your bike's pads. Yes, you could run them to the absolute limits of their service, but you'd be gambling that you won't want to use your brakes multiple times in a series of corners or won't need maximum braking power. So, plan on replacing your pads when a minimum of 2 mm of the pad material remains. (Even with 2 mm of pad material left, your braking power could be compromised with heavy use such as stopping from highway speeds while riding two-up.)

The directions for this project are based on changing front brake pads, but the steps are the same for the rear. (For drum brakes, see Project 8.) Begin by placing your bike on a bike lift or on its sidestand in gear. Follow these instructions on one caliper at a time. Unbolt the caliper from the fork leg and remove it from the rotor. Now remove the brake pads. On some calipers, you may need to pry the pads apart with a large flathead screwdriver before you can slip out the pads. Some calipers secure the pads with a retaining pin and spring clip that you will need to remove. If the pin is a screw-in type, loosen it before removing the caliper.

Slip out the used pads, but before you set them aside, inspect them closely. Did they wear evenly? If one pad is thinner than the other, take a look at the pistons to see if they may be sticking. Another symptom of a dragging caliper piston is having one end of the pad more worn than the other. These signal the need for a caliper rebuild.

After you have the pads out, don't just toss in another set and be done with it. You need to clean the caliper components and the disc's swept surface. Cleaning the caliper removes the obvious stuff such as brake dust, but it also plays an important role in preventing the need for rebuilds. Most brake fluids absorb moisture through the piston seals into the sealed environment of the hydraulic system. And that's not all. That moisture combines with the brake dust and grit and whatever else happens to be wedged into the calipers, forming a kind of shellac that bonds to the surface of the pistons and keeps them from sliding smoothly through the rubber seals.

Since your goal is to clean the caliper and its pistons, exposing more of the pistons' surface would be a good idea. Impatient mechanics may choose to squeeze the brake lever slightly to push the pistons out of their

Once you've freed the caliper from its mount, you may want to pry the pads apart with a screwdriver. Then they'll slip out of the center as shown here.

bores. However, squeeze too much, and a piston will pop out of the caliper, creating a huge mess. Smart folks slip a thin piece of wood in between the pistons to prevent the dreaded "piston pop." Be sure to leave enough room for a toothbrush to clean the pistons.

Some mechanics rely on contact cleaner to flush the contaminants out of the calipers. Others arm themselves with Simple Green or other organic cleaners to cut through the mung. Who's right? According to the brake manufacturers, petroleum-based cleaners such as contact cleaner swell the rubber O-rings that seal the calipers. You should only use organic cleansers to flush out the calipers. You have been warned.

Using the toothbrush and organic cleaner, carefully rub the shellac off the pistons. The side of the pistons toward the top of the caliper is the hardest to reach and may require some creative dexterity to get to it. If you see any signs of leaking brake fluid, go buy the caliper rebuild kit. You should also pay attention to the surface of the pistons. If you see rust or pitting, prepare for a rebuild. Once the calipers are clean, rinse with water and blow them dry with compressed air to keep any of the solvents from being carried into the brake fluid as you slide the pistons back into the caliper.

While the pistons are extended out in their cleaning position, you won't be able to install the new brake pads. Press the pistons back into the caliper one or two at a time. Be sure to hold the other pistons in place while you do so to keep them from popping out from the fluid pressure. If the pistons refuse to press fully into their bores (so that they are almost flush with the caliper), check the master cylinder's reservoir. Since the fluid in the caliper is displaced back into the reservoir, fluid that was added as the

With the organic cleaner of your choice, spray the surface of the piston to loosen the gunk. Don't be stingy; soak the caliper. Placing a pan under the caliper will help you catch the mess.

Use a soft-bristled brush to help remove the dust buildup. A brass brush can be particularly helpful with tough deposits.

level dropped in the reservoir may hydraulically lock the system if it completely fills the reservoir. Suck some fluid out with a syringe and continue to press the caliper pistons into place.

To finish your caliper cleansing, on single-action calipers (those with pistons on only one side of the caliper), pull back the rubber cover protecting the pins upon which the caliper slides back and forth. Using the same organic cleaners, remove the old, tired grease and check for any notches or unseemly wear to the pins. Lube the clean pins with waterproof, high-temperature grease and slide them into position under their protective covers.

Before you install the new pads, you need to prepare the disc's swept surface. Brake pads leave traces of material in the pores of the rotor's surface. If you're changing pad compounds, removing this material is vital if you want the new pads to bed in properly. However, you still want to remove any buildup of pad material on the disc face, using a piece of 60- or 80-grit sandpaper, even if you're replacing the old pads with ones of the same compound. With your bike on a lift or jack with the wheel off the ground, simply press the sandpaper against the disc and spin the wheel a few times. If the front wheel is still on the ground, have an assistant push the bike while you apply the sandpaper. Repeat

After the caliper's interior is clean, press the pistons back into their bores. Otherwise, you won't be able to slip the new, thicker pads into place.

If you can't get the pistons back into their bores far enough to allow the disc to slide between the pads, you can use a pry bar to gently open up more space. Be careful not to mar the pads' surfaces.

the process on the other side of the disc. Remember: You're not trying to score the disk, just remove the buildup. Finish the disc prep by spraying it with brake cleaner and wiping it down with a clean rag.

Now you can replace the pads. Pay special attention to placing the factory anti-squeal plates or springs in the proper position. Check the other caliper if you can't remember how the pieces go together. If the pads are not held in place with a pin, pay attention to the way they fit in the caliper, and verify that they haven't shifted once you've remounted the caliper to its bracket. Torque the fasteners to the proper specification. Also, pump up the brake lever as soon as you finish remounting the caliper to build pressure and seat the pads against the discs. This way you won't experience the unfortunate surprise of having the lever come all the way back to the grip as you try to stop at the end of your driveway.

After you've mounted your new pads, you need to break them in to receive the maximum stopping power they have to offer. If you go out and immediately hammer them repeatedly, you run the risk of glazing the surface of the pads and vastly reducing their coefficient of friction and performance. Read the "Brake Pad Break-In" sidebar for the best way to ensure that your brakes give you the best they can.

Cleaning any pad residue off the disc with sandpaper will speed the bedding in of your new binders. This step is vital if you've changed pad compounds.

Every time you change your brake pads, you should also clean and lube the slider on a single-action caliper to ensure even pad wear.

Brake Pad Break-In

When you put new pads on your bike, don't rush out and try maximum braking on them right away. The surfaces of both the pads and the discs aren't really as smooth as they look. In fact, they're made up of lots of hills and valleys—they look more like a saw blade up close—as in, under a microscope. The break-in procedure wears down the pad so that the hills and valleys match each other, giving the maximum surface-area contact. Get too aggressive too soon, and the hills melt and glaze over, lowering the coefficient of friction for the pads and reducing braking performance.

When installing your new pads, you cleaned the disc with brake cleaner, right? Oil, grease, or brake fluid can ruin pads. Begin pad break-in by riding in a parking lot and lightly applying the brake, bringing the bike slowly to a stop. Do this a few times with slightly more pressure each time. Now, go for a ride. Just like when you have new tires and you know not to immediately throw your bike into corners, allow yourself some extra room for braking around town. Begin to vary your braking pressure. Each stop should be more firm than the last. Allow for some cooling between applications (i.e., no maximum-braking exercises).

There is no magic number for how long it takes to break in a set of pads. If you vary the pressure and don't build it up too quickly, a short ride may be enough. However, if you're ever in doubt about how to break in a set of pads, simply look at the instructions that came with them—the manufacturers know.

PROJECT 8

Drum Brake Maintenance

Time: 30 seconds to 1 hour

Tools: Wrenches, sockets, Allen keys, torque wrench, dial caliper, rags, brake/contact cleaner, high-temperature grease, bike lift (optional)

Talent: ★

Tab: $

Parts: Brake, shoes, brake shoe springs (if not included with shoes)

Tip: To keep from breathing hazardous materials, don't blow off the brake panel with compressed air

Benefit: Predictable rear brake performance

Although the number of cruisers shipping with drum rear brakes has dropped over the years, enough of them are still around to warrant inclusion in this book. Being mechanical devices, drum brakes are subject to wear and tear, which shouldn't be surprising when you consider that they convert motion into heat through friction—just like those fancy discs on most bikes.

Drum brakes require a little regular maintenance that should take you all of 30 seconds every month or so. Since the brake pedal is more-or-less physically attached to the brake shoes (unlike hydraulic systems), the pedal's freeplay will get greater as the shoes wear. A quick press of the pedal with your hand will give you all the information you need about the state of the freeplay. From the pedal's resting position, measure the distance the pedal is depressed until you feel the shoes contact the drum. If that measurement is greater than the factory spec, you'll want to tighten the nut at the drum end of the brake rod. In many cases, you won't even need a wrench to do this. Simply press the cam lever (the lever attached to both the drum brake assembly—or brake panel—and the brake rod) slightly against the spring holding it into position against the adjuster. Since the adjuster usually has a machined curve to match the clevis pin in the lever, you'll be able to make changes in half-turn increments. Once the freeplay is back to spec, you're done.

When you're adjusting the pedal freeplay, take the time to note the wear indicator when you take up the freeplay. The manufacturers were nice enough to build these indicators into the outside of the brake panel since you can't visually check the condition of the brake shoes without removing the wheel and disassembling the drum brake. As the indicator gets closer to the end of its range, the thinning material of the brake shoes is vulnerable to the same issues as front brake pads. Plan on replacing the shoes when they're about 80 percent used up to be sure that maximum power will be available to you.

Every two years or 10,000 miles, you should plan on disassembling the brake panel and lubing the moving parts. This would be a good time to clean out the built-up brake dust and give the assembly a close visual inspection. Of course, you should also perform these chores when replacing the shoes.

Before jacking (or lifting) your bike's rear wheel off the ground, bikes with chain drive (seen often on many drum brake–equipped cruisers) need to have the adjuster's locknuts loosened so that there is enough slack in the chain to allow the wheel to be removed. Then give each of the adjusters three full

Don't forget to remove the bolt securing the drum to the torque arm. You'd be amazed how many people are momentarily surprised when the rear wheel won't come off after the axle is removed.

Carefully measure the thickness of both shoes to make sure that they don't need to be replaced.

turns of slack and retighten the locknuts to maintain their adjustment relative to each other. (Once you get the wheel back on the bike, you'll just tighten the adjusters those three turns for a properly adjusted chain.) Owners of bikes with shaft drives should check with their factory service manual for rear wheel removal tips. Different bikes have different requirements. For example, the Yamaha V-Star 650 requires that you pull the shaft itself free with the rear wheel.

Remove the bolt connecting the torque arm to the drum brake assembly and the nut from the end of the brake rod. Now, loosen the axle nut and spin the axle nut off the axle. Jack up the bike so that the wheel is about an inch off the ground. While holding the wheel in position (use your foot to take the strain off the axle as you slide it free), remove the axle. Now, push the wheel forward to create some slack in the chain. Remove the chain from the sprocket and hang it on the swingarm. The wheel can now be lowered out of the swingarm. You may need to jack the bike up farther to help the wheel clear the fender.

Place the wheel on the floor with the drum brake side facing up and wiggle the brake panel free. You'll see lots of brake dust and grime on the inside of the panel, but resist the urge to blow it out with compressed air. Those particles can contain some pretty toxic stuff. Instead, clean the parts (in a well-ventilated area) with brake cleaner. It will flush the dust away without taking it airborne. Don't worry that it'll also remove the grease from its rightful place. You're going to reapply it anyway. If you decide to clean the inside of the drum, take care not to get the cleaner in the wheel bearings.

With the brake panel exterior on a surface that won't scratch it, measure the thickness of the pads with a dial caliper to make sure they are within specification. As with disc brake pads, most brake shoe failure takes place on the last 20 percent of the shoe material. So, if the shoe thickness is less than 3 mm, replace the set. (The Vulcan 800 factory service manual, for example, specifies a service limit of 2.6 mm.) You should also check the internal diameter of the drum itself with your dial caliper.

Grasp the brake shoes with a clean rag (to keep the shoe surface clean in case you're just lubing the assembly and not replacing the shoes, too) and lift them free of the assembly. The shoes will fold toward the center of the panel as they pivot around the anchor pin and the cam shaft. Be careful, the springs holding the shoes against the pin and cam are fairly strong. Set the shoes aside and wipe the panel clean of any crud.

Protect your hands from the shoe material and the shoes from grease by gripping the two parts with a clean rag as you rotate them free of the retaining pin and cam.

Turn the brake panel over and, using a Sharpie, mark the brake cam lever's position by drawing a line from the gap in the lever across the shaft to the other side of the lever. Unscrew the pinch bolt and remove the lever. Now, slide the camshaft out of the panel. Set the wear indicator aside. Clean the shaft and its associated hole in the panel and check for wear. If you notice any wear, consult your factory service manual for tolerances to make sure the parts are still within specification.

Once everything is clean, lube the camshaft and the mounting hole in the brake panel with high-temperature grease. Slide the shaft into the panel and rotate it until the flats on the cam are pointing toward the center of the panel. Place the indicator on the shaft with its pointer at the right end of the useable range marker on the panel. Line up the Sharpie marks on the shaft and the lever and remount the lever. Tighten the bolt to spec. Apply high-temperature grease to the contact surfaces of the anchor pin and the cam. Add a little dab of grease to the brake shoes' spring-mounting holes. (While Honda thoughtfully includes new springs with the new brake shoes, you'll need to buy them separately for most—if not all—other cruisers.) Since you don't want to gum up your new shoes (or old ones, either), be sparing in your application of grease.

Before returning the brake panel to its home inside the wheel, rotate the cam lever to make sure that everything is working. Check that the wear indicator is pointing at the beginning of its range (or where it was pointing prior to disassembly if you just cleaned and lubed the unit). If not, remove the cam lever and adjust the indicator. When you're ready to put the wheel back in the swingarm, begin by cleaning and greasing the axle. Carefully slide the wheel into the swingarm and lift it up into the forward position you used to remove the chain. Take the chain from the swingarm and put it back on the sprocket. Pull the wheel backward until the hole in the hub lines up with the hole in the swingarm. (Having the axle already hanging in the swingarm eases the next step.) Slide the axle into the hub far enough to support the wheel. Move to the opposite side of the bike where you can see through the swingarm to the hub. Lift the wheel until the hole lines up with the swingarm. Tap the axle through using your hand or a dead-blow hammer.

For tips on how to remount your wheel on a shaft-driven cruiser, see Project 13, Installing Aftermarket Wheels.

Mark the cam lever to ease returning the lever to its proper position after disassembly.

Snug down the axle nut but not so tight that the chain adjusters can't move it. Make sure the wheel is all the way forward against the adjusters by giving the rear of the tire a couple of whacks with your dead blow hammer. Reset the chain adjusters the three turns you loosened them. If you haven't adjusted your chain in awhile, now is a good time. Verify that the marks on the chain adjusters match up on both sides. Coat the clevis with a light application of grease, slip it into the cam lever, and slide the brake rod into place. Set the adjuster into roughly the same position on the rod it was in prior to disassembly. Loosely (finger tight) attach the torque arm to the brake drum assembly. Before torquing down the axle nut, spin the rear wheel and lock of the rear brake by pressing quickly on the pedal. With the brake still applied, tighten down the bolt securing the torque arm. Now, torque the axle to factory spec. Make sure that the chain adjusters are snugged up and their locknuts are tight. Verify that everything is in place and the wheel turns freely. Adjust the brake pedal freeplay to factory settings. If you've installed new shoes, follow the break-in instructions in Project 7.

Apply a coating of grease to the pin and cam lobes as well as the spring mounts. Note how the coating covers the parts without leaving much extra grease.

PROJECT 9

Hydraulic Fluid Change

Time: 45 minutes

Tools: Wrenches, sockets, rags, torque wrench (inch-pounds), Teflon tape, brake bleeding tool (optional), clear hose, glass jar for brake fluid, rags

Talent: ★

Tab: $

Parts: Brake fluid

Tip: Only use fresh fluid from unopened containers, and never mix fluid brands.

Benefit: Better lever feel, no nasty surprises

Complimentary Modifications: Add braided-steel brake lines (see Project 10)

Maybe this weekend, when you needed to do a panic stop because a dog ran out in front of you, your brake lever came much closer to the grip than usual. Or maybe you just checked your brake (or hydraulic clutch) fluid and noticed that it has turned dark or cloudy. Or you could just be following your bike's maintenance schedule (good for you!) and the time has come to freshen your hydraulic fluid. However you came to this point, you should know that fresh hydraulic fluid is vital for proper performance.

Most hydraulic fluids have a taste for water and will gradually suck moisture past the rubber seals in your calipers. If your fluid is contaminated with water, heavy brake use will raise the temperature to the point where the water will boil (at a significantly lower level than pure fluid would), leading to brake fade and the dread-inducing, lever-to-the-grip experience.

Cruiser owners should swap out hydraulic fluid at least once a season. Some people claim you should do this as part of the winterizing process, so that the moisture-free fluid sits in your bike all winter. Having never had problems develop while my bike was in storage, I prefer to flush the fluid at the beginning of the season so that I get the benefit of fresh fluid in the spring. Those whose priority is maximum braking performance will most likely want to follow this schedule, too.

The tools required for this project are minimal, but you can buy some specialized tools if you like—you know you want to. The Mityvac bleeding system (from Lockhart Phillips) is perfect for bone-dry systems—such as when you've installed stainless-steel lines. If the system is already primed, an old jar and clear hose will work just fine. The fluid you add to your system should come from an unopened container. Remember the hydraulic fluid's thirst for water. Although DOT 5 fluids are silicon-based and, therefore, don't absorb moisture, some riders don't like the feel at the lever as much. Most OEs still recommend DOT 4 fluid. Also, make sure you buy a name-brand fluid. Buy more fluid than you think you need. (If you're having trouble with bubbles in your lines, the best cure is to run lots of fluid through the system to sweep them out.) Finally, never mix brands of brake fluids. If you need to top off the reservoir and don't know what type or brand of fluid is already in the system, flush the entire system to avoid any problems of interaction between the different

When wrapping the bleed screw's threads with Teflon tape, make sure you wind it so that screwing in the valve tightens the tape. Also, be careful not to cover the bleed holes.

This Mityvac is a handy tool for bleeding hydraulic systems—particularly bone-dry ones. Be prepared to empty the catch container at least once when freshening fluid. Since you'll be working by the caliper instead of the master cylinder, don't forget to keep an eye on the fluid level in the reservoir.

formulations, even though—in theory—all DOT 4 (and DOT 5.1) fluids should play well together.

Begin with your bike on its sidestand or a lift. Since brake fluid damages paint and some powder coats, remove any vulnerable parts or cover them with rags. A preparatory step required for those using vacuum bleeders, and optional (but recommended) for all others, is to wrap the threads of the calipers' bleeder valves with Teflon tape. Vacuum bleeders create so much suction that they will draw air into the system past the bleeder valves' threads, making it impossible to tell when all air has been removed from the system. Teflon tape fills the minuscule gaps between the threads when the valve is not completely closed and is a worthwhile modification to all calipers. One warning, though: Don't just remove the valve with the caliper mounted to the bike. Since the master cylinder is higher than the caliper, fluid will leak out all over the place, ruining your brake pads. Raise the caliper equal to or above the reservoir to keep the mess to a minimum. Of course, if your lines are empty—as when you've just added stainless-steel ones—this is not a problem.

Rear calipers have their own idiosyncrasies to consider. Since the brake line travels horizontally from the master cylinder to the caliper, air bubbles get trapped in the line more easily. To assist the bubbles in their travels, unmount the caliper and hold it higher than the master cylinder, allowing the bubbles' tendency to rise to keep them moving as you run the new fluid through the system. If you're pumping the pedal, make sure you place a spacer between the brake pads to keep the piston(s) from popping out of the caliper.

When using a vacuum bleeder, you can save some time by sucking the excess fluid out of the master cylinder. Then wipe any visible grit out of the reservoir. If you're not using a vacuum bleeder, don't worry about running the old fluid through the system—unless you see dirt or other visible impurities. Don't add fresh fluid to the reservoir until most of the old fluid has been pumped out of it and into the brake system itself.

Swapping the fluid in the system is pretty easy. Put a box-end wrench on the caliper's bleeder valve, then press a length of clear hose over the nipple. This must have a snug

Keeping a big arch in the line will keep bubbles from being drawn back into the caliper. Don't be stingy with your fresh fluid. Keep pumping it through the system even after you think you're done. Some bubbles are tenacious.

fit. The other end of the hose should empty into a container that is capable of standing on its own. (Glass containers, while breakable, are usually heavy enough to stay put unlike plastic bottles or aluminum cans.) Slipping the hose through a hole in the screw-on cap is ideal. Don't forget to leave a breather hole in the lid, or the built-up pressure could pump the fluid out of the hose when disconnected from the nipple, creating another mess. (Don't ask me how I know this.)

Pump the lever a few times and hold the last squeeze. Using the wrench, open the bleed valve until the pressure forces fluid into the hose. Hold the lever all the way in until you've closed the valve. Repeat this process several times until you have a couple inches of fluid in the hose directly above the bleeder. Now, open the valve slightly—only until you can just squeeze fluid out with the lever and no more. Continue to squeeze and release the lever until you see the fresh—usually clearer—fluid emerge from the bleeder. A short pause between every squeeze and release will make sure that any bubbles expelled from the system don't get sucked back in when you release the lever. Pay special attention to the reservoir while you are pumping the fluid through the system. If you let it run low, you'll introduce more air in the system and have to start the bleeding process

When topping off the reservoir, don't fill beyond the full mark. Be sure to clean the diaphragm that floats on top of the fluid before installing it. Always use name-brand hydraulic fluid.

over from scratch. If you started with empty lines, plan on running several more ounces of fluid through the system after you stop seeing bubbles to make sure that none remain.

When you're sure that the hydraulic system is free of old fluid and/or bubbles, slowly squeeze the lever as you tighten the bleeder valve. Pump the lever several times and hold. Open the bleed valve to release the fluid. Do this several times. You'd be surprised how often one last bubble pops out during these final, high-pressure bleeds. If one does, run a few more ounces of fluid through the system. You will be rewarded with firmer lever feel. Torque the bleed valve to the recommended spec and finish by topping off the reservoir. Don't forget the other caliper! Make sure that the rubber diaphragm on the master cylinder is clean before placing it on top of the hydraulic fluid and tightening down the reservoir cover. Wipe away any traces of the brake fluid before it has a chance to damage any painted surfaces. Finally, as with all waste fluids from your bike, be a good citizen and recycle the old stuff.

Those of you with vacuum bleeders shouldn't feel left out. All the steps are the same, except that the fluid will be sucked from the bottom end of the system instead of forced from the master cylinder. The same cautions apply, though. Accidentally let the reservoir go dry, and you'll need to start over. Just keep pumping the bleeder to maintain a constant but not excessive pressure. When no more bubbles present themselves, you're done.

Any discussion of hydraulic system bleeding usually includes a debate about which method is better: the master cylinder push-through or the vacuum tool suck-through. While many mechanics will subscribe to either one or the other, I've found a combination of the two works best. When bleeding dry lines, nothing gets them primed quicker than a vacuum bleeder. However, when swapping fluid, I prefer the pump-through method. So, when installing new lines, I use both methods. First, draw the fluid through with a suction tool, followed by the final flushing of air with the master cylinder method. Regardless of which technique you incorporate, the goal is to fill the system completely with fresh hydraulic fluid and no air bubbles. You'll be glad you took the time to do this the next time you grab a handful of brake.

PROJECT 10 Stainless-Steel Brake Line Installation

Time: 1-2 hours

Tools: Wrenches, sockets, bike lift (optional), rags, torque wrench (inch-pounds), Teflon tape, brake bleeding tool (optional), clear hose, glass jar for brake fluid, zip ties, crush washers

Talent: ★★

Tab: $ to $$

Parts: Goodridge brake lines, brake fluid

Tip: If you've modified your bike by raising or lowering the bars or extending the swingarm, order a custom-length kit.

Benefit: A firmer squeeze at the lever, better feedback, and better braking performance.

Complementary Modifications: Replace brake pads (see Project 7)

Although you can't see OE brake lines expand when you squeeze the lever like you could in the Bad Old Days, fitting a set of braided, stainless-steel brake lines to your cruiser can have a dramatic effect on its stopping power. The initial onset of braking will be much quicker since stainless lines don't expand at all. And since the lines are sheathed in metal (usually with a protective plastic outer coating), you don't have to worry about stainless lines cracking from age and exposure to the sun. Also, the Teflon interior is less prone to becoming brittle than rubber. So, a trip to the aftermarket will give you better braking and longer-lasting lines, to boot. Oh yeah—and they look cool next to your polished fork and chrome caliper covers.

Most of the major line manufacturers such as Goodridge have premeasured kits available for almost every cruiser manufactured in the last 10 years. So you shouldn't have any problem finding one for your ride. However, if you've modified your bike by raising or lowering the bars or extending the swingarm, you'll probably want to have a custom-length kit special ordered for you. Some manufacturers offer build-'em-yourself kits where you cut the lines and attach the fittings. While these kits are great for custom applications, you need to order each individual part, right down to the angle of the bend on each banjo fitting, so be forewarned.

Before you begin installation, check to make sure that all the lines in your kit are the correct length. Nothing will make you crazier than having a line end up an inch short while your bike sits idle with the entire system disassembled.

Quickly attaching the new line to the stocker will tell you if it is the right length. This will also help you to prepare for any idiosyncrasies of your bike's brake line routing.

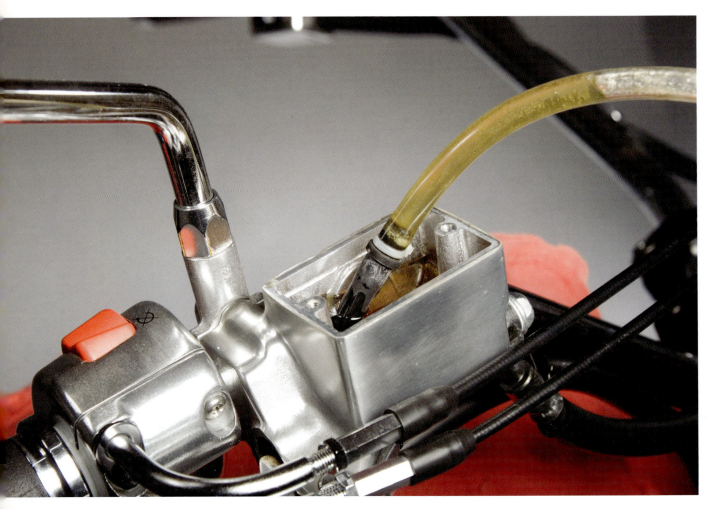

Sucking the fluid out of the reservoir will speed up the task of draining the system. Place the cap back on the system—but don't screw it down—to keep dust out of it while you're changing the lines.

You're then stuck with either no ride until you get the correct part or reinstalling lines you want to take off anyway. The simplest way to check the length of the lines is by routing them loosely in place and zip tying them to the existing lines. Although this takes a couple of extra minutes, you can tell right away if the lines will have the proper amount of slack in them. Having your brake lines go taut before the fork is fully extended would be a Very Bad Thing.

Even in the best-case scenario, changing hydraulic lines is messy. Since brake fluid can damage paint and other shiny stuff on your bike, you should remove or cover any vulnerable painted surfaces. You will also want to get the old system as empty of fluid as possible before removing the lines. A vacuum bleeder is ideal for this. Begin by sucking the extra fluid out of the reservoir. Then attach the hose to a caliper's bleeder valve. Give the bleeder a couple of pumps to build up the suction and crack the valve until fluid starts to be drawn into the catch tank. Keep pumping until the system is dry. Do this for both front calipers. The rear caliper follows similar steps, but remember the bleeding issues noted in the previous project.

Unscrew or unclasp all the fasteners holding the hydraulic line in place. Using a ratchet, remove the banjo bolt from the caliper. To keep the fluid leakage to a minimum, wrap the banjo with a rag and secure it with a zip tie or piece of tape. Remove the master cylinder banjo and feed the line out of the chassis. Now, feed the new line into place following the exact same route as the stock line. Some aftermarket front brake kits will use two lines from the master cylinder instead of a T-junction farther down the line. Be sure you run the correct line to each caliper. (One is usually longer than the other.)

Always replace the crush washers when you remove the banjo bolts. The soft copper (for steel banjos) or aluminum (for aluminum banjos) is designed to conform to any irregularities on the fitting or mounting surface. A washer should be used on both sides of the banjo. If two banjos are being bolted together (as on the front brake master cylinder), be sure to use a crush washer between the two banjos, as well. Screw the banjo bolts in finger tight and check your hose routing before you torque things down. You don't want any sharp bends or kinks in the lines. If things don't line up right, you may have the banjos at the wrong mounting point. Hydraulic line manufacturers spend a lot of time making sure that the fittings have the same bend as the OE lines they replace—if something doesn't look right, it probably isn't.

Once you've torqued banjo bolts down, make sure you attach the lines to the chassis at all the original points. Sometimes you'll need to use zip ties to hold the thinner, stainless lines to the OE clips. Although most stainless lines are sold in protective sheaths, bare, braided stainless-steel lines can cut through metal like a hacksaw. If your lines are uncoated, make sure you wrap the lines with tape or spiral wrap specifically designed for the purpose at all potential points of contact with the chassis. You can find cool chrome wraps in the aftermarket. Then, follow the bleeding instructions in Project 9.

Sure, they look great—but just wait until you feel the difference. (Don't grab too hard the first time, though.)

Notice the lack of curves as the stainless line leaves the caliper. If it had the wrong bend in the banjo fitting, the line could kink.

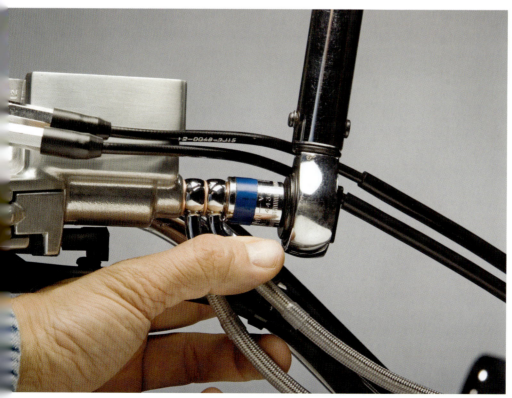

The OE system only used a single line at the master cylinder, so the second aftermarket line needs to be held in place while the banjo is torqued. Note the three crush washers used to seal the system.

PROJECT 11 | Caliper and Master Cylinder Rebuild

Time: 2-3 hours

Tools: Wrenches, sockets, rags, torque wrenches (inch-pounds and foot-pounds), Teflon tape, brake bleeding tool (optional), clear hose, glass jar for brake fluid, compressed air, pick, Scotch-Brite, Craftsman soft jaws pliers covers, latex gloves, small thin piece of wood, high temperature grease, crush washers

Talent: ★★★

Tab: <$ to $$

Parts: Piston seals, dust covers, pistons and calipers as required, master cylinder or piston as required, crush washers

Tip: Be extremely careful when clearing the bleed holes because they're delicate

Benefit: Brakes that don't drag

Complementary Modifications: Replace pads (see Project 7), upgrade brake lines (see Project 10)

If you changed the brake pads (Project 7), you also looked for uneven wear on the brake pads as a sign of sticky brakes. Well, even if no pistons in the calipers are sticking, you should consider rebuilding a caliper if you can't remove the brake dust deposits from a piston with solvent and a brush. Proper functioning of the brake system depends on pistons that move easily within their bores. Calipers aren't the only part of the brake system that will need to be rebuilt occasionally. The master cylinder, which powers the entire hydraulic system, needs attention from time to time, too. If it has a problem, you won't get full performance from your brakes. The most common issue is a piston with a worn primary or secondary cup, which allows hydraulic fluid to flow past the seal, resulting in low power or leaking at the brake lever. The master cylinder also has several tiny orifices that, if they clog, can prevent proper operation of the brakes.

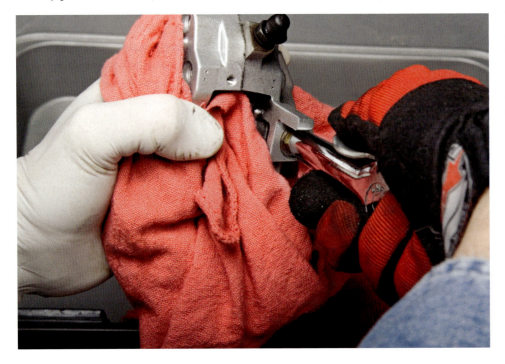

When using air pressure to push out the pistons, take it easy. A little pressure can provide a lot of piston movement. Keep your fingers clear and use a rag to contain the pistons.

If a piston is stuck, try removing it with soft jaws and a set of adjustable pliers. Also, try rotating it in the bore as you attempt to pull it out.

Calipers

No matter how many pistons they have, calipers are broken into two categories: single-action and dual-action. Single-action calipers have pistons that only press from one side of the disc and are most common in rear brakes. Dual-action calipers squeeze the disc with pistons from both sides. Rebuilding these two kinds of binders follows exactly the same steps—you just do more on dual-action units.

The rebuild requires you to press the pistons out of the calipers. The recommended method is to drain the system and carefully blow out the pistons with compressed air. If you don't have compressed air, you can use the hydraulic fluid and lever to achieve the same result. Unfortunately, this technique is so messy that most people only try it once. Be sure to place a container under the caliper to catch the fluid. Follow the same procedures for restraining the pistons described below. And don't forget latex gloves to protect your skin from the hydraulic fluid.

Begin the rebuild by draining the system and removing the caliper. To avoid a mess, wrap the line's fitting with a rag and tape it in place. Remove the pads, springs, clips, piston insulators, and any other hardware around the caliper's pistons. It's easiest to remove pistons from single-action calipers. For a one piston, single-action caliper, place a rag between the piston and the bracket for the second brake pad.

Carefully blow compressed air into the hole for the banjo fitting. Keep the air pressure low and your fingers clear of the piston! If the caliper has more than one piston, you will want to place a piece of wood a little thicker than your disc, as well as the rag, into the caliper. This will keep one piston from popping out and leaving the other(s) deeply recessed in its/their bore(s).

Similarly, one-piece, dual-action calipers should have their pistons blown out with a piece of wood in between the sets of pistons. The goal here is that the wood is thin enough to allow the pistons to fully extend, but not pop free. Two-piece, dual-action calipers can sometimes be unbolted, giving you better access to one set of pistons at a time. (Check your factory service manual before attempting this. Some manufacturers recommend against it.) Since you'd have to fashion a means of controlling the pistons as you blow them out of their bores, I'd recommend treating the calipers like one-piece units until you have the pistons extended.

If you're lucky, you'll be able to pull the pistons free of their bores with your fingers. If not, use a set of "soft jaws" on a pair of adjustable pliers. These plastic covers will keep the metal of the pliers from touching the piston and scratching its surface while still giving you a remarkably strong grip on it. After the pistons are removed, use a pick to gently pry out the dust covers and inner seals. Be careful

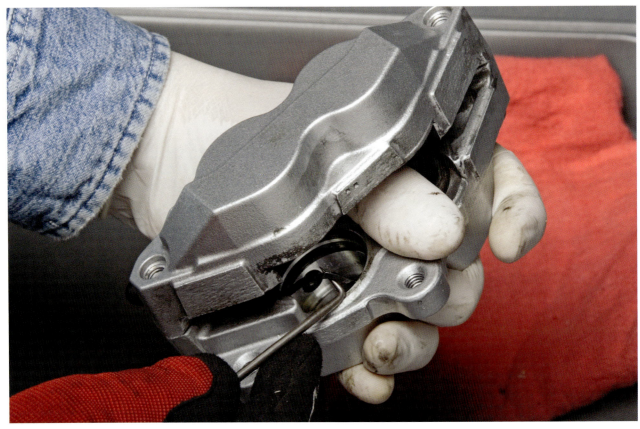

Be careful not to damage the piston bore as you remove the seals.

not to scratch the piston bores—although doing so is less of a disaster than with the pistons. Thoroughly clean the interior and exterior of the calipers and pistons with brake cleaner or an organic solvent. If the parts are particularly dirty, soak them for a few hours in an organic cleaner. Remove any traces of the solvent and blow the parts dry, making sure that all fluid passages are clear.

Inspect the pistons. If you find any rust or pitting, polish the surface with Scotch-Brite until it is perfectly smooth. If the corrosion is too great, replace the piston. Similarly, buff away any pitting in the piston bores, and replace the caliper if you can't make the interior look like new. Spray the parts with brake cleaner and blow them dry again.

Although you don't need to replace the O-ring piston seals unless they're damaged, why would you go through all this trouble and not replace them? Install fresh seals and the dust seals by applying a coating of fresh brake fluid and slipping them into place. (Note: The dust seal is closest to the bore's opening.) To avoid damaging the seals, don't use any tools to install them—just your fingers. Apply a coat of brake fluid to the pistons and slide them all the way into their chambers. If you disassembled the caliper halves, install new O-rings and torque them together to factory specs.

Remove any corrosion with Scotch-Brite. If you can't restore the piston to a polished surface, replace it.

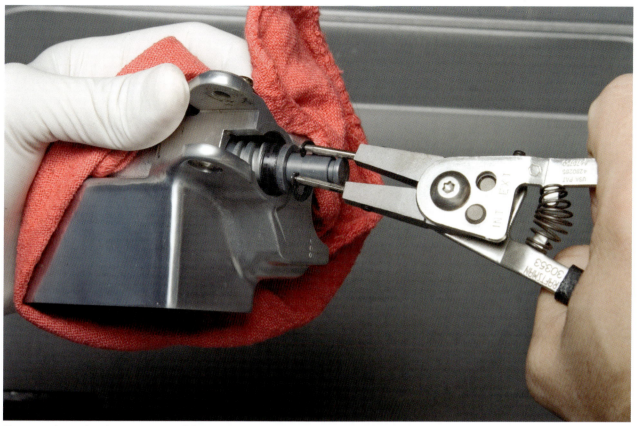

Once you have the dust cover out of the way, press the piston in slightly and remove the circlip.

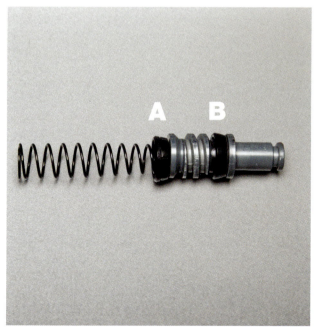

Inspect the primary (A) and secondary (B) cups for any signs of swelling or wear. The piston should be free of corrosion.

Install any anti-rattle springs, clips, and other parts you removed from the caliper. On single-action calipers, apply a coating of high-temperature, water-resistant grease to the caliper holder shafts and the holder holes. If the boot and dust cover for the shaft is cracked, you should replace them. Now is also a good time to wrap the caliper bleed-valve threads with Teflon tape to make bleeding the system easier (see Project 9). Install fresh pads and bolt the caliper into place. Torque the banjo fitting and its fresh crush washers to spec.

Master Cylinder

Drain the system and remove the brake lines. On units with a remote reservoir, remove the hose connecting the reservoir to the master cylinder and unbolt the reservoir from its mounting bracket. Remove the brake lever or pedal, and brake switch wires. Take the master cylinder off of its mount for the remainder of the steps. Using a pick, carefully remove the dust cover from the master cylinder and piston and inspect it for cracks or punctures. If you find any, replace it. Look inside the piston bore for any signs of brake fluid leakage. If you see any, the piston needs to be replaced. Remove the circlip that holds the piston in its bore. The piston should slide out far enough for you to remove it by hand. Do not

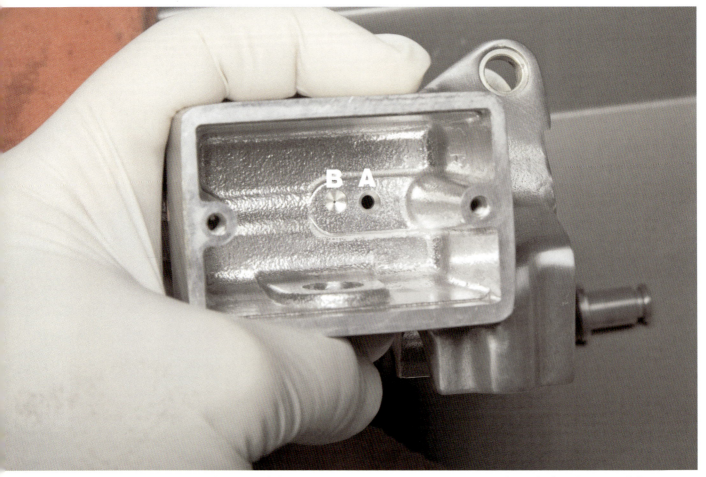

Visually inspect the supply (A) and relief (B) ports in the master cylinder. If compressed air won't clear them, carefully use a piece of wire. The same ports reside in master cylinders with remote reservoirs, but the space is much tighter.

attempt to remove either the primary or secondary cups from the piston, as you will damage them. The steps are the same for both front and rear brake master cylinders.

With the piston removed, give the master cylinder a thorough flush with an organic cleaner. (The rubber parts on the piston require special treatment and should only be cleaned with brake fluid or isopropyl alcohol.) Make sure all solvents are rinsed away with brake cleaner from the master cylinder internals. Blow it dry with compressed air. Closely inspect the piston bore. If you find any corrosion, rust pitting, or scratches, replace the master cylinder and the piston (which may have had its cups damaged by the abrasive corrosion). Make sure the supply and relief ports between the reservoir and the piston bore are clear. A piece of grit blocking the relief port can cause your brakes to drag, reducing pad and disc life. Blow the ports clear with compressed air. If you can't clear the port with air, carefully use a thin piece of wire, but beware, damaging the port will affect the function of your brakes.

A tight seal is necessary for maximum braking power. So give the piston a once-over, or even a twice-over. If either cup is worn, rotted, or swollen, replace the entire piston assembly. If the piston has any visible rust or corrosion, replace it. Finally, check the return spring for any kinks or corrosion and replace if necessary.

Before reassembling your master cylinder, coat the interior of the piston bore with brake fluid. Coat the piston cups with fluid. Make sure the return spring has the correct orientation before inserting the piston. Press in the piston as you reinstall the circlip. Don't forget to make sure that the dust cover is properly installed, sealing the piston from the outside environment. Remount the brake lever, and be sure to lube the pivot with grease and torque the pivot bolt's locknut. Bolt the master cylinder to the handlebar or frame. Attach the brake line with a fresh set of crush washers and torque the banjo bolt to spec. Reconnect the reservoir and bleed the system as in Project 9.

PROJECT 12

Aftermarket Caliper and Disc Installation

Time: 1-2 hours

Tools: Wrenches, sockets, Allen key sockets, torque wrench, thread lock, hammer, eye protection, bike lift (optional), impact wrench (optional)

Talent: ★★

Tab: $$ to $$$$$ (Complete systems cost big bucks.)

Parts: Aftermarket disc, aftermarket calipers, new pads, DOT-4 brake fluid.

Tip: Rap the disc mounting bolts a couple of times to free the thread lock.

Benefit: More stopping power and better styling.

Complementary Modifications: Add braided stainless-steel lines (see Project 10)

Although OEM cruiser brake technology usually provides excellent stopping capability, you may find that you want to replace your discs and/or calipers for a variety of reasons. The good news is that aftermarket discs and calipers often work and look better than the stockers. If you shop around, you can often find more than one option for replacement binders. The bad news is that none of the parts come cheap.

What you get for your money is pretty impressive, though. Many front discs are now floating models, meaning the swept area of the disc is loosely mounted on a carrier. This space between the two pieces allows the disc to expand without warping in high-temperature situations. Since the swept area and the carrier now have unrelated jobs, their composition can vary. While the disc may be polished or plain stainless steel, the carriers can vary from anodized aluminum to polished aluminum to chromed steel. You have a multitude of options available to you. Also, if none of those choices appeal to you, remove the disc from the carrier, send it out for the cosmetic treatment you desire, and remount the disc. The same is true of calipers. You can send the stockers out for chroming, polishing, or powder coating. If you want a different style caliper or a sportier set with, say, six pistons, you merely need to check the aftermarket for a bolt-on kit. If a prefab kit isn't available, you can have a machine shop create some caliper mounts to fit your new caliper and disc combination.

Mounting a New Disc

Installing new discs has few pitfalls. You begin by loosening the front axle and caliper bolts so that you're not really cranking on them while the wheel is off the ground. While most cruisers will have an axle and axle nut that you can

On dual-disc front wheels, place the wheel on a spare tire to keep from pressing on the bottom disc while you wrench on the top one. Once all the bolts have been loosened, a speed wrench makes quick work of removing them.

Some discs were designed with a particular rotational direction in mind. Inspect the disc closely for an arrow that indicates the proper wheel rotation. (If you're unsure, check the arrow on the tire and match that.)

loosen with sockets, some cruisers have started following a trend that has been popular in sportbikes for a while. In this case, the axle is recessed in the fork leg and requires a large hex key to remove it. Fortunately, the manufacturers haven't moved to the obscenely large hex keys that are almost impossible to find . . . yet. Using a breaker bar, loosen the axle about one turn. Jack the wheel about an inch off the ground. Next, unmount and hang the calipers by something other than their hoses (like a zip tie). While supporting the wheel slightly with your foot, slide the axle free. Lower the front wheel out of the fork. While you may be tempted to lay the wheel on its side so you can remove the discs, be careful not to support the wheel with the opposite disc. It can't withstand lateral forces and will bend easily. Instead, place an old tire (or old two-by-fours) under the wheel.

People sometimes run into trouble on what seems like the easiest step. The manufacturers often use thread lock on the bolts securing the disc, and these bolts usually incorporate Allen (hex) heads. So if you're not using an impact wrench to loosen the bolts, insert an Allen socket and extender into each bolt and give it a couple of good raps with a hammer. Don't forget eye protection. The loosened thread lock should allow the bolts to spin out easily. For stubborn bolts, use a bit of heat on the area around the bolt—just enough with a propane torch to melt the thread lock. While you've got the disc off, you might want to clean the parts of the wheel that are hard to get to otherwise.

Some discs are designed to be mounted to either the left or right side of the wheel so that they rotate the correct way against the pads. Read the instructions for your new discs to make sure. Also, some will have the rotational direction marked on the disc or carrier. This is vitally important. Place the new disc in position, put thread lock on the bolt threads, and screw them into place so that they are finger tight. Next, torque them down in a crisscross pattern in a two-step process (first half the valve on all the bolts, then to the final torque) to ensure that the disc mounts evenly. Before moving on to the next disc, take a rag moistened with contact cleaner and wipe down the swept surface of the disc. Frequently, discs will be shipped with a coating of protective oil. You want to be certain that this is completely removed or it will contaminate the pads and dramatically reduce your braking power. If the disc has a particularly heavy coat of oil, you can wash and blow dry the disc before you mount it. Remember, you must use new pads with your disc.

Aftermarket Calipers

While most front calipers on stock cruisers feature four pistons, you can order aftermarket kits with six or more pistons. Why do you want these? More pistons will press on a larger portion of the disc, multiplying the effective force transferred from your fingers to the pads. Even if you already have four-piston

Make sure you wipe off the protective oil coating on the disc before remounting the wheel. In fact you should always wipe the discs before you position the calipers. It's far too easy to leave grease or some other contamination on the swept surface while you're wrenching. Be sure to wear gloves when using contact cleaner.

calipers on your bike, you can still find four-piston models that deliver better braking than the stockers. Besides, performance brake calipers for cruisers are usually more visually pleasing than those that OEMs must create to a price point.

For such a pricey modification, installation is pretty straightforward. If you've ever changed the brake fluid, lines, or pads, you've done about half of the work before! Swapping the stock brake system for an aftermarket one is a two-part process. First, remove the old system. Second, mount the new one. Well, duh.

Begin with your bike on a lift with the wheels off the ground. Remove the front wheel and swap the rotors. Drain the system as described in Project 10. An empty system is much easier to disassemble. Since you really should install braided steel lines with the new calipers, take note of the path taken by the lines before removing the stock system (see Project 10). You'll also want to clean all the parts before you pack them away for storage (or for sale on eBay). Your calipers will most likely include spacers required to mount them on your bike in the proper orientation to the discs. Simply bolt the spacer to the OE caliper mounts. Slide the caliper into position on the spacer and torque it down.

If you're venturing into the unknown by buying calipers without the prebuilt spacers necessary for your bike, consult a local machine shop that specializes in building parts for racers. They'll have the experience required to measure and create the proper spacers for you. Then get them polished or chromed.

Once the whole kit is torqued into place, just install the brake lines and bleed the system as outlined in Project 9. Finish by bedding in the pads according to the manufacturer's instructions. Then go show off your trick new parts and improved street cred.

While the original caliper usually mounts directly onto the fork leg, aftermarket units often require a spacer to position the caliper relative to the disc. *Photo courtesy of Performance Machine*

PROJECT 13 | Installing Aftermarket Wheels

Time: 2 hours

Tools: Wrenches, sockets, front and rear stands, rags, torque wrench, measuring tape, Loctite

Talent: ★★

Tab: $$$$$

Parts: Aftermarket wheels

Tip: Check spacers and disc attachment points to be sure they match *before* you mount the tires.

Benefit: Great looks! Most billet wheels are heavier than stock and will actually diminish performance.

Complementary Modification: Install new brake pads (see Project 7) or mount an aftermarket brake system (see Project 10)

In its most basic sense, installing a set of aftermarket wheels is not much different than remounting the stock rims after a tire change or brake disc swap—except for the extra money and the envy of your riding buddies. Seriously though, mounting up a set of shiny, sexy aluminum wheels carries a performance cost you should note. At the time of this book's publishing, no wheel manufacturer makes aftermarket cruiser wheels that are lighter than stock. Some make ones that are just about the same weight as the stockers, though. The simple truth is: Heavier wheels slow acceleration, handling, and steering.

The theory behind this claim is straightforward. In motorcycle performance, weight is everything. The OEs produce some reasonably light wheels. Why? Because any reduction in unsprung weight (weight not supported by the suspension) makes it easier for your suspenders to help the tires track across pavement irregularities. So even saving a couple of pounds here is a big deal. Next, consider the weight of the wheel in regard to acceleration. As Kevin Cameron says in his excellent *Sportbike Performance Handbook*, "A pound saved in a wheel rim . . . is worth two pounds anywhere else on the machine. A wheel has to be accelerated twice; once in a straight-line, and also in the second sense of rotating around its own center." Since wheels rotate, generating gyroscopic forces, a lighter wheel will turn in quicker and accept steering inputs more readily. Riders who like flicking their bikes into turns will notice the effect heavier wheels will have on steering. Fortunately, the wide bars on cruisers give you plenty of leverage with which to muscle the front wheel.

When choosing your new wheels' finish, you should note that one trade-off with polished wheels versus chromed ones is their propensity for oxidizing. However, you'll need to clean and wax both chromed and polished aluminum wheels more often than the coated stockers—particularly if you live near the ocean. You may be wondering why the selection of aftermarket wheels isn't as great for metric cruisers as it is for SAE ones. Well, the shaft drive interface is the issue here. Each cruiser manufacturer has its own design for the shaft housing, and many models within a manufacturer's line are different. This requires a separate wheel design for each model. Some aftermarket companies have designed rear wheels around a removable hub, making it possible for a model-specific hub to fit several wheel styles. Perhaps Yamaha had the best approach with the Road Star by making the hubs' specs match those of the SAE wheels, thus opening the door to those multitudes of Harley-friendly wheels in the aftermarket. A fairly recent development in the aftermarket is to create a universal hub for the interface between the shaft drive and Harley-spec rear wheels. Honda Direct Line has done this for VTXs. Expect more companies to follow this trend.

When taking delivery of aftermarket wheels, be sure to measure a few dimensions before mounting them up. If something doesn't fit, you'd have a harder time exchanging a wheel if you scratched it mounting the tire. While you don't have to measure to aerospace tolerances, try to be as accurate as possible. Measure the distance between the mounting surface of the disc carriers on dual front discs. Also, measure the distance to the outside of both wheel spacers when they are fully seated in the hub. Compare these numbers to those of your stock hoops. A millimeter or two difference is probably your margin of error if you're using a tape measure. If you get a discrepancy of any more than that, remeasure both sets of

One nice thing about belt drives is that you can mount pulleys that match the style of your wheel. You can also vary the size of the pulley to change gearing to the wheel.

wheels. Otherwise, you could mount the wheel off-center and run into problems such as the caliper resting against the disc when you torque the axle nut. Since wheel manufacturers often sell the same wheels for multiple bike models and ship them with spacer kits, receiving a wheel with the wrong kit is not uncommon. Call the seller if you have any questions about the dimensions.

Front Wheel
Since the axle nut is torqued on pretty tight, loosen it while the front tire still rests on the ground. To prepare to loosen the axle, loosen the axle clamp bolts at the bottom of each fork leg. While most cruisers will have an axle and axle nut that you can loosen with sockets, some cruisers have started following a trend that has been popular in sportbikes for a while. In this case, the axle is recessed in the fork leg and requires a large hex key to remove it. Fortunately, the manufacturers haven't moved to the obscenely large hex keys that are almost impossible to find . . . yet. Using a breaker bar, loosen the axle about one turn. On virtually all bikes, you'll have to remove the calipers before you remove the wheel. Even on those that don't require it, remounting the wheel will be much easier if you do. So, loosen the caliper mounting bolts while the wheel is on the ground and remove the calipers, too. Don't forget to support the calipers with something other than the brake lines. Now, you can use your jack or stands or lift to raise the front wheel off the ground. Spin the axle free and remove the wheel.

Have a tire shop experienced in handling billet wheels mount your chosen rubber on the new wheel (otherwise you risk having them scratched). Swap the discs or install the new discs as described in Project 12. Installing the front wheel is the reverse of removal—with a couple additional steps. Wipe the axle clean and apply a thin coat of high-temperature grease to it. Before you torque the axle, you'll want to snug it up and take the bike off the front stand. Remount and tighten the calipers. Pump the brakes up, squeeze the brake lever, and hold it while you bounce the front end a few times to make sure that the wheel is centered in the fork. Torque the axle, pinch bolts, and caliper bolts to factory specifications.

Rear Wheel
Since cruiser final drives come in essentially two categories—belt and shaft—this section will be divided into two parts. (Yes, some cruisers do have chain drive, but I have never seen one with billet wheels that wasn't a one-off project built by accomplished customizers.) Still, some portions of the modification remain the same. You will need to either support the bike on a lift or a bike jack. If you choose a jack, make sure that the jack is placed far enough rearward to keep the front tire on the ground. As a safety precaution, wrap the front brake lever with a bungee cord or a couple turns of tape to keep your cruiser from rolling forward off the jack. If you're using a lift, make sure that the bike is strapped to it securely since you'll be muscling the wheel around in the swingarm. However, before you raise the rear wheel, don't forget to loosen the rear axle nut. Raise the rear of the bike until the rear tire is about an inch high off the ground.

Shaft Drive
Completely remove the axle nut. Place your foot under the tire and lift up with your toes to take the weight slightly off the axle. When you've got the right pressure under the wheel, the axle will easily slide free. Rotate the rear caliper carrier away from the hub and support it on the bike with something other than the brake line (a zip tie usually works). Next, shift the wheel slightly to the right to free the splines

Measuring the distance between the brake mounts will help you make sure that your new wheel is correctly set up for your cruiser.

on the hub off of those on the drive housing. Once the wheel is resting on the ground, you may have to jack the bike up farther, depending on what kind of rear fender your bike has. Before you install the new wheel, wipe all the old grease off the splines and give them a liberal coating of fresh grease. Grease the splines on the new hub and the axle, too. Roll the wheel into position in the swingarm and lower the bike until it is at the height that will only require you to lift the wheel about an inch to mount it to the drive.

Getting the splines on the hub and the drive can be a challenge. First, the teeth have to line up. Second, the wheel needs to be perpendicular to the housing to fit in place. One thing that can help is to lift the wheel into its approximate position and slip the axle through the hub and out the other side of the swingarm. This will help keep it perpendicular to the drive. With the bike now supporting the wheel, you can slowly rotate the wheel until splines line up and allow the hub to shift into place. Using a flathead screwdriver, pry the rear brake pads apart. Pull the axle out until it just barely goes into the hub. Rotate the caliper carrier into place, slide the axle through the carrier and swingarm, and tighten down the axle nut. Torque it once the rear wheel is back on the ground.

Belt Drive
While belt drives offer many advantages, like light weight (when compared to shaft drives), quiet operation, and relative ease for altering the final drive's gearing, that flexibility (pardon the pun) comes at a slight cost. One slight bother with belt drives is that removing and mounting wheels takes a bit more effort. Also, you need to check the belt's tension every 2,500 miles or so. If it needs adjustment, you need to set aside 20 minutes to adjust it.

To remove the wheel, begin by unbolting any belt covers that look like they will interfere with wheel removal. (Got that factory service manual?) Remove the brake caliper on a Road Star/Warrior from its mounting bracket. If your belt is properly adjusted and you're not changing the size of either pulley, you can save some reinstallation time by loosening the locknuts on the belt adjusters (hold the adjuster in place with a wrench while loosening the locknut), loosening the caliper mount (on the Road Star/Warrior), adjusting the pulley three full turns of the adjuster looser on the bike (on the Road Star/Warrior you'll be turning the adjusters on either side of the bike in opposite directions, so pay attention), and retightening the locknuts to hold the adjusters in place.

Completely remove the axle nut. Place your foot under the tire and lift up with your toes to take the weight slightly off the axle. When you've got the right pressure under the wheel, the axle will easily slide free. Slide the wheel forward to create slack in the belt. Slip the belt off the pulley and push it out of the way. If you didn't remove the caliper, rotate the caliper carrier away from the hub and support it on the bike with something other than the brake line (a zip tie usually works). Once the wheel is resting on the ground, you may have to jack the bike up farther, depending on what kind of rear fender your bike has. Transfer the disc and pulley to the new wheels if necessary.

When you're ready to put the wheel back in the swingarm, begin by spreading the brake pads with a flat-head screwdriver. Next, carefully slide the wheel into the swingarm and lift it up in the forward position you used to remove the belt, and remount it. If the caliper stayed on its bracket, lift the wheel into position to slip the disc between the pads as you hold the bracket in place on the swingarm. You may need to try a couple of different angles to get the disc in the caliper. Once the disc is in the caliper and the belt is on the pulley, pull the wheel backward until the hole in the hub lines up with the hole in the swingarm. (Having the axle hanging in the swingarm eases the next step.) Slide the axle into the hub far enough to support the wheel. Move to the opposite side of the bike where you can see through the swingarm to the hub. Lift the wheel until the hole lines up with the swingarm. Tap the axle through using your hand or a dead-blow hammer.

Snug down the axle nut. Reset the belt adjusters the three turns you loosened them. If you haven't adjusted your belt in awhile, now is a good time. Verify that the marks on the adjusters match up on both sides before torquing down the axle nut.

To check your belt's adjustment, you need either a belt tension gauge or a fishing scale that measures 10 pounds. Push the belt up with 10 pounds of pressure and look at the measurement guide in the belt cover that most belt-driven cruisers have. When you need to adjust your belt, follow this precaution: Loosen the axle nut just enough to enable the belt adjusters to move it. If the axle nut is loosened too much, you can accidentally knock the rear wheel out of alignment.

If the belt is too loose, tighten the adjusters an equal amount to keep the rear wheel in alignment. If you have to go quite a bit, make quarter-turn adjustments, switching to eighth-turn adjustments as you get close to your goal. Measure the slack after every change. If the belt is too tight, loosen the adjusters two full turns and use a rubber mallet or dead-blow

If you get a strange reading when measuring the hub, make sure that the spacers are bottomed out in the holes and are sitting straight. Yes, you should take this measurement before mounting the brake disc and tire.

Look at all those teeth! No wonder it can sometimes be a challenge to get your wheel to mate with the shaft drive. Don't forget to apply fresh grease.

hammer to knock the rear wheel forward against the adjusters and begin again. When the slack is within factory specifications, torque the axle nut. Next, tighten each adjuster about one-eighth of a turn against the axle. Hold the adjuster in position and set the locknuts. Road Star and Warrior owners should also remember to retorque the caliper bracket (and caliper mounting bolts if you didn't do that when you reinstalled the caliper). One last time, verify that the marks on the adjusters match up on both sides of the swingarm.

Finish up this project by making sure that everything is in place and the wheel turns freely. Pump up the rear brake lever to make sure the pads aren't dragging. Finally, add a fresh cotter pin to the axle nut. OK, now quit polishing those beautiful new hoops and get out there and show 'em off!

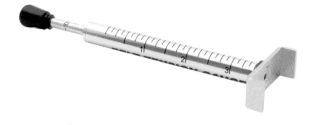

Measuring belt slack just got easier with the release of this Motion Pro belt tension tool. *Photo courtesy of Motion Pro*

With the hub hanging on the axle, lift the opposite side of the wheel until the holes line up. Then push the axle home.

Tightening the axle nuts to the proper torque specification is vital for your safety. Cranking them down too tight can compress bearing spacers and lead to bearing failure. Leaving them too loose can let them spin free with disastrous results.

BRAKES

SECTION 4
SUSPENSION

Projects 14–21

PROJECT 14 | Adjusting Preload

by Eric Goor

Time: 1 hour

Tools: Two assistants, riding gear, metric tape measure, open-ended wrench or socket, shock preload adjusting tool or long screwdriver and hammer, hacksaw or pipe/tubing cutter (for PVC spacers), jack

Talent: ★

Tab: $

Parts: PVC pipe to cut spacers

Tip: Adjust the preload to suit the load the bike will carry—rider, passenger, gear, etc.

Benefit: With preload adjusted for proper sag, your suspension will be able to do its job better.

Springs are pretty simple-minded. The harder you push on them; the harder they push back. Preload adjusters on motorcycles use this relationship to help riders fine-tune their suspension to fit the variety of loads their motorcycles are asked to carry. Riding solo on a twisty road puts an entirely different load on the suspension than riding two-up with soft luggage on the same road. In order for the suspension to have its full range of travel to soak up pavement irregularities, you'll want to take advantage of the rear preload adjusters most cruisers have. If you invest some time, you can find both the loaded and solo preload settings for your bike, which makes changing them before a ride as simple as turning a few wrenches. Some of the bigger, touring-oriented cruisers will also have air pressure adjusters to set front preload, too. Even forks that don't have external adjusters can have their preload altered. It just takes a little more time.

Measuring Sag

Before you decide to move beyond your bike's factory suspension settings, you must do two things. First, make sure that all the components of the system are in good working condition. Fix any leaks and perform any other maintenance chores before making modifications. If you don't, you'll likely have to start from scratch when you finish the maintenance in the near future. The second thing, which in reality is the first step toward setting your suspension to your riding style, is to set your bike's sag by adjusting the suspension's preload.

Proper sag is important for two reasons. First, a suspension unit needs a certain amount of room within its range of travel to work properly. If you have too little sag, your bike will be prone to "topping out" the suspension as it extends to its upper limit. Similarly, too much sag could allow you to experience the unpleasant jolt of "bottoming out." Second, once your sag is set, you will be able to ascertain whether or not your suspenders' spring rate is correct for your weight and size.

Your bike's sag is broken into two categories: "static sag," which is the distance your bike compresses its suspension from fully extended when you and your riding gear are aboard, and "free sag," which is the distance your bike settles from full extension under its own weight. Using the algebra you suffered through in high school and an equation popularized by Race Tech founder Paul Thede, you'll be able to determine whether you need to increase or decrease your preload to reach the magic combination.

In order to measure the static sag, you'll need two assistants. A metric tape measure will also make the calculations easier than an SAE one. Before you mount the bike, you'll want

This Mean Streak makes measuring sag easy. If your bike has fork covers that prevent access to the stanchion, measure the sag from the fender to the bottom of the triple clamp.

With the rider in position, extend the suspension and let it settle slowly back into its sag. Next, press down to let it rise up into place. The average of those two measurements will give the true reading.

The process of measuring suspension is the same for the rear suspension, except that you need to measure to a spot directly above the axle.

to measure the suspension when it's completely topped out. For the fork, lift on the grips until the front wheel begins to come off the ground. On traditional forks, measure from the stanchion wiper to the bottom of the triple clamp. (If your bike has fork covers, you'll need to measure from the top of the fender to the bottom of the triple clamp. Also, make sure you measure from the same location every time.) Measure from the wiper to the top of the axle clamp on inverted forks (like those found on the Yamaha Warrior and Kawasaki Mean Streak). We will name this number L1 and write down the measurement.

Now it's time to measure the suspension while you're on the bike. Have one of your assistants hold the bike from the rear while you get into your riding position. (If you're setting sag for riding two-up or with a full load of luggage, be sure to have these on the bike for this measurement.) Your other assistant should push down on the fork and let it slowly rise up until it stops. The new measurement will be called L2. The front end should now be lifted and allowed to settle slowly down until it stops, forming measurement L3. Exactly in the middle of measurements L2 and L3 is the point where the fork would want to live in a frictionless system. So, the average between the measurements would be (L2+L3)/2.

Armed with this information, you can determine the static sag by subtracting the average measurement calculated

SUSPENSION

above from L1, or to write it out as an equation: static sag = L1−(L2+L3)/2. (Aren't you glad you stayed awake in class that day?) Now that you've got this number, what does it mean? Suspension gurus generally agree that between 30 to 35 mm (1.2 to 1.5 inches) is optimum sag. If you have too much sag, you'll need to increase the fork's preload. Conversely, if you have too little, back off on the preload a bit. See your factory service manual for instructions on how to do this, if you're lucky enough to have preload adjusters (if not, see below). The jockey-sized and big-boned members of the audience may be wondering if they should fiddle with these figures to account for their mass. In a word, no. The static sag figures are an accepted constant. Measuring the bike's free sag will reveal if any alterations need to be made for rider size.

Once you have the front suspension dialed in as described below, repeat the measurement process with the rear suspension. The key to getting accurate measurements out back is to pick a solid point on the frame or bodywork directly above the axle. If you don't measure straight up from the axle, you may get inaccurate numbers.

Changing Preload
If you want to adjust the preload on a fork that doesn't have factory adjusters (and most cruisers don't), you can create custom spacers by cutting them out of the largest PVC pipe that will fit inside the fork. Although you don't have to take the front end off the ground, removing the pressure from the front suspension will make reassembly easier. When changing spacers without jacking the front end up, only remove the cap from one side of the fork at a time, or the bike will crash down to the bottom of the fork travel. You have been warned.

If your bike has air adjustable preload (front or rear), all you need to change the preload is this nifty pump from Progressive Suspension. *Photo courtesy Progressive Suspension*

Begin with your bike on a jack or a lift. While a jack isn't required (you can support the front of your bike with a couple tie downs thrown over a garage rafter), the job is much easier with solid support under the chassis. If you don't support the front end, the bike will drop on the fork as you remove the cap of each tube, leaving you no recourse but to lift the front end when you want to put the cap back on.

Next, remove any parts that may interfere with your access to the top of the fork legs, such as windshields and handlebars (which can be carefully laid on a soft cloth on the tank without removing the cables or hoses). Loosen the bolt on one side of the top triple clamp to relieve pressure on the cap's threads. Remove the fork cap either by unscrewing it with a wrench or by (the increasingly rare

Once you remove the fork cap, you'll see the top of the spacer. Any length you add to your new spacer will make it that much harder to reseat the fork cap.

Notice how the locking ring has been loosened from the adjusting ring. The black mark will make it easy to keep track of how far the ring has been turned.

method of) pressing down on the cap and removing the circlip. The cap is under pressure from the spring, so be prepared for it to pop out. Remove the stock spacer and measure its length. Since the amount of sag you need to gain/lose is almost a one-to-one ratio to the amount you need to remove/add to the spacer, you can easily approximate the right length with simple arithmetic. While you've got the saw out, cut a couple sets of PVC spacers in quarter-inch increments on either side of your calculated length. File down any rough edges on the PVC and clean the spacers of any grit. Label the spacers with a Sharpie. Slip the new spacer into place with any washers you may have also removed with the stock spacer. If you've increased the preload, expect to work a bit to get the fork cap in place. Don't forget to torque the screw-on caps and the triple clamp's pinch bolts back to spec.

Changing rear shock preload is fairly easy on all bikes. Most stock shocks will have either a stepped adjuster or threaded, locking-ring adjusters. The stepped adjusters usually ramp the preload over five or six settings. Using the tool supplied in the factory toolkit, simply lever the collar onto the proper step. The process should take less time than it took to get the tool out from under the seat. Then just remeasure the sag to see if you reached your goal. Locking ring adjusters can also be altered with a tool. Motion Pro makes a clawed adjuster that mounts to a 3/8-inch ratchet. Many aftermarket shocks also ship with an adjuster. If you don't have a tool, a long screwdriver and hammer will work in a pinch.

Begin by loosening the locking ring (the one farthest away from the spring). Using a Sharpie or a scribe, mark the adjusting ring so that you can count the number of turns you increased/decreased the preload. The fine-pitched threads move the adjusting ring approximately 1 mm per revolution. When you're roughing in the preload, make adjustments in full-turn increments. Fine-tuning the preload will be done with much smaller increments. Once the preload is set, tighten the lock ring down to keep the adjusting ring from backing out. Don't jam the lock ring down more than a quarter turn, or you may have trouble loosening it next time you adjust the preload.

Now that the static sag is set, you can measure the free sag to make sure your bike has the correct rate springs. Measure the amount the bike sags under its own weight. If you want to be really anal, you can use the equation above, but a single, quick measurement will tell you if your spring rate is in the right zip code. The free sag should be between 0 and 5 mm. Simply lifting the weight of your bike to see if it moves up slightly before topping out the suspension will give you an idea of how much free sag it has. While this may not seem to make sense, if your suspension has no free sag, your spring rate is too soft. The soft rate forced you to use too much preload to get the desired sag. If you have a bunch of free sag, your spring is too stiff—see Project 15 to learn how to install new fork springs. Changing the rear spring will require a trip to your local suspension guru or aftermarket company.

A final note about preload: If you take the time to measure the preload required for solo riding and two-up, write down the settings. When you need to change the preload for different riding situations, all you'll have to do is take out the wrenches and make the changes to the shock. (Since most cruisers don't have fork preload adjusters, set the preload for the kind of riding you do most and leave it.) Not having to lure assistants with free beer to measure the sag will save you time *and* money.

While changing the preload doesn't get any easier than using a stepped adjuster, you can't fine-tune the preload the same way you can with a threaded collar.

Make sure your cuts on the PVC spacers are square. A miter box or pipe cutter will help with this. Also, cut, clean up, and label three or four spacers to ease the swapping process as you dial in the front end.

PROJECT 15 | Installing Fork Springs

Time: 1 hour

Tools: Wrenches and/or sockets, torque wrench, circlip pliers or jeweler's screwdrivers (for non-screw-on caps), press (for non-screw-on caps), claw-type pick-up tool to grab cartridge piston rod (optional), saw with miter box or pipe cutter, rags, jack or front stand

Talent: ★★ to ★★★

Tab: $

Parts: Fork springs, PVC tubing

Tip: Progressive-rate springs vary their rate, depending on the amount they are compressed, for better handling of the bumps

Benefit: Better, more comfortable ride

Complementary Modification: Change fork oil (see Project 16)

So, you set your sag only to find that your spring rate was wrong. Or perhaps your stock springs are starting to sack after a couple years of riding. Either way, you're looking at installing a new set of fork springs. Swapping fork springs in traditional forks, be they damping rod or cartridge units, is pretty dang easy. Inverted forks may require special tools and techniques for the swap (see Project 19), but the process is still pretty easy.

Start by lifting the front end with the jack or lift until just before the tire leaves the ground. On bikes with air adjustable preload, release all the pressure by pressing the pin in the center of the valve. If the handlebar on your cruiser will prevent you from being able to lift the spring straight out of the fork leg, remove the bar from its clamps and lay it on the tank. (Of course, you should protect the tank with a towel for padding. Always protect the tank when wrenching on the triple clamp.) Loosen the bolt(s) securing the top triple clamp to the fork leg. Bikes with screw-on caps only need to have the caps unscrewed with a wrench or deep socket. Caps secured with a circlip or retaining ring must be pressed in to take the pressure off of the circlip. Two tools can make this much easier. Some automotive part pullers will hook over the triple clamp and press in the cap via a thumb screw. A woodworker's corner clamp can achieve the same result for much less money (usually around $5). Press down on the cap just enough to take the strain off the circlip. Using a jeweler's screwdriver or pick, remove the circlip, and slowly ease the cap out with the press. You will want to hold a rag over the cap, as it may jump out from the force of the spring.

Damping-rod forks are the easiest to swap springs but only slightly more so than cartridge units. Once the cap is removed, pull out the spacer and any washers and lift out the spring. Turning the spring counterclockwise as you lift it out helps to free it of excess oil, but you'll still want to wipe it off and place it on a clean rag.

Folks with cartridge forks need to go through one or more additional steps to free the spring and spacers: You will need to disconnect the fork cap from the piston rod by loosening the locknut securing it to the cap with a wrench on the nut and socket on the preload adjuster. Then simply unscrew the cap from the rod and remove the spacer, spring, and washers. (Note their order for reassembly.)

When you disassemble a cartridge fork on a cruiser, you may notice something funny about the fork's internals. Yep, often only one of the legs will have a cartridge. The other one will only have oil and a spring. When you consult your factory service manual, you may also discover that the factory has different requirements for oil volume and height within the leg.

Although you may be tempted to drop your new springs into place, don't—unless you want to wipe up the fork oil that splashes out. If you are using progressive rate springs that are wound more tightly at one end than the other, some manuals will recommend placing the spring with the tightly wound end down. Why? According to the folks at Progressive Suspension, the direction of the spring wind makes no mechanical difference, though sometimes orienting the spring this way will lessen the spring noise.

To keep your bike clean, slowly pull the spring out of the fork leg. Turning it counterclockwise can also help leave the oil in the fork.

Racers, on the other hand, recommend keeping the tightly wound section up to make it part of the bike's sprung weight (the part of the bike supported by the suspension) instead of the unsprung weight on the wheel that must track over pavement irregularities. Either way, the spring will work the same—you decide.

If you need to make your own preload spacer to fit this spring, use the largest PVC that will fit inside the fork leg. You will need to cut the spacer perpendicular to the tubing. A miter box or pipe cutter will help, but isn't necessary. Those with non-adjustable forks will want to make a variety of spacers in quarter-inch increments on either side of the spring manufacturer's recommendation to aid in the setting of sag. Use a knife and a bit of sandpaper to deburr the spacers. Wipe both the interior and the exterior of the spacers clean before installing them. Don't forget to write

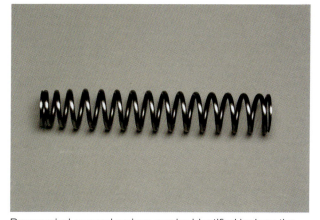

Progressively wound springs can be identified by how the wind is tighter at one end. This helps the spring to function at one rate fully extended and a stiffer one as it compresses.

SUSPENSION

Although you can remove a circlip fork cap by pressing down on the cap with a screwdriver while simultaneously removing the circlip, this cheap woodworking clamp makes the job much easier.

their length on them with a Sharpie, so you don't have to measure them each time you change them. Make sure that the preload spacer is the length specified by the spring manufacturer before placing it, and any necessary washers, into the fork leg. Reassemble the freshened fork in the reverse of the way you disassembled it.

The rare cruiser fork with rebound adjusters requires special assembly techniques, so consult your factory service manual for the recommended method. To finish up your fork, make sure you torque the fork caps and don't forget the pinch bolts on the triple clamp. Once you've reset your sag, you should feel a noticeable improvement in your suspension.

These precut and labeled spacers will speed the setting of your bike's preload if you don't have adjustable forks.

PROJECT 16 | Changing Fork Oil

Time: 1-2 hours

Tools: Wrenches, sockets, torque wrench, screwdrivers, Allen keys, metric tape measure, jack, circlip pliers or jeweler's screwdrivers (for non-screw-on caps), press (for non-screw-on caps), caliper or ruler, Ratio Rite or another graduated container, spring compression tool (for inverted forks)

Talent: ★ to ★★★

Tab: $

Parts: Fork oil

Tip: Overfill the fork slightly if you are using a suction-type tool to set oil height.

Benefit: Consistent damping from year to year

Fork oil must be one of the most neglected components on any motorcycle. Unless an owner follows the recommended maintenance interval to the letter or has a quality shop perform maintenance, fork oil may go years without being refreshed. I've seen oil come out of forks that looks like it's been in there since the well-used machine was set up at the factory. Fork oil, like motor oil, loses viscosity over time. If ignored, the fork will cease to perform properly and internal components such as the slider bushings will begin to wear. So, even if you're not planning to upgrade your front suspension, be sure to replace your fork's slippery stuff every two years or 15,000 miles, or at the interval your factory service manual recommends.

Changing the fork oil begins by removing the fork from the chassis. Remove the front wheel as outlined in Project 12. Remove the front fender by unbolting it from the fork legs. Caps secured with circlips should be removed while the fork leg is on the bike. Screw-on caps should only be loosened, but not removed, before the leg is removed from the triple clamp. On bikes with air adjustable preload, release all the pressure by pressing the pin in the center of the valve. Loosen the bolt(s) securing the top triple clamp to the fork leg. Cruisers without fork covers only require that you loosen the bolts on the upper and lower triple clamp and slide the fork free.

Bikes with fork covers require that the fork covers be removed, too. Remove the top triple clamp (with the bar still attached) by unscrewing the bolt or nut securing the top clamp to the steering stem. Wiggle the bar front to back as you walk the triple clamp free. Once it slips off, lay the bar on a thick piece of padding on the tank. Now unscrew anything securing the fork cover to the lower triple clamp and follow the above directions to extricate the fork legs from the lower triple clamp. (Note: I've seen people remove fork legs without bothering with the fork covers, and remounting the fork is much more work with the covers still in place than simply taking the dang covers off in the first place.)

Screw caps should be removed with a socket or wrench. Holding a rag securely over the fork cap is good advice—even for those with screw-type caps. As you reach the last thread, the cap will tend to fly off. If you're not holding on to it, you could be injured, or worse, the cap could get dinged up.

Damping rod forks are the easiest to prepare for changing oil, but not by much. Once the cap is removed, pull out the spacer and any washers and lift out the spring.

After you remove the front wheel, unbolt the fender. You may need to squeeze the sides slightly to remove it. Beware, the paint scratches very easily.

SUSPENSION

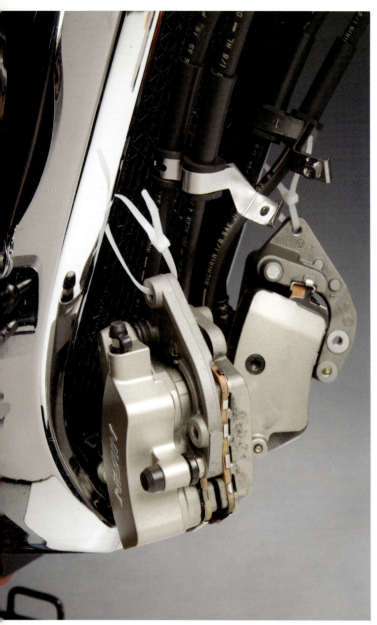
Whenever you do any work that requires removing the calipers, don't leave them hanging by the brake lines.

Turning the spring counterclockwise as you lift it out helps to free it of excess oil, but you'll still want to wipe it off and place it on a clean rag. Folks with cartridge forks (including inverted forks, such as those on the Mean Streak) need to go through a couple additional steps to free the spring and spacers: You will need to disconnect the fork cap from the cartridge piston rod by loosening the lock-nut securing it to the cap with a wrench on the nut and socket on the cap. Next, remove the nut from the piston rod to free up the spacer, spring, and washers. (Note their order for reassembly.)

If you can't see the nut at the base of the fork cap, your bike requires a cartridge fork compression tool (available as a factory service part from your bike's manufacturer, or from an aftermarket company such as Race Tech) to compress the fork spring to reveal the locknut securing the cartridge to the cap. Once the preload spacer is compressed enough to reveal the nut, use a pair of wrenches, loosen the locknut, and spin the cap off the cartridge piston rod.

Empty the oil into a suitable container for transport to a recycling center. In order to make sure that all the old, dirty oil is expelled from the fork, you'll need to pump the slider up and down a minimum of 10 times. Cartridge forks (both standard and inverted) require that the piston rod be pumped to expel the oil. If you can't grip the locknut on the end of the rod while pumping it, you can buy the screw-on factory part intended for this job or call Race Tech.

Measure out the amount of oil recommended by your factory service manual before adding it to the fork. If you plan on using a fork oil level tool, add between half an ounce and an ounce of extra oil. If you're using the dipstick method, follow the manufacturer's recommended amount. Pour the oil into the fork and fill the system by pumping the fork and piston rod a minimum of 10 times. Keep pumping until you no longer hear air being expelled.

The fork oil height should be measured with the fork fully compressed and no spring installed. Although you can measure the oil height with a simple dip stick fashioned out of a coat hanger (as I did for many years), buy one of the fork oil level tools sold by Race Tech or Progressive Suspension if you plan on changing your fork oil more than once. You'll be glad you did. These tools feature a metal tube with an adjuster on the end that you measure to the appropriate oil height and lock in place. Then, when you place the end of the tool on the fork, it sucks out all the excess oil and stops at the specified height. If you're using the dip-stick method, you'll measure the height, add a little (or subtract a little), and remeasure until you reach the proper level.

Don't worry about torquing fork caps into the fork tubes until the unit is reinstalled in the triple clamp with the lower clamp bolts torqued to spec. Before you tighten the triple clamp pinch, however, make sure that the top of the fork tube (not the cap that sits on top of it) is flush with the top of the triple clamp (that is, unless your factory service manual specifies otherwise). Folks with fork covers can save this step until after the second leg is mounted. Just don't forget to do this final tweaking once the top triple clamp is in place and torqued to the steering stem.

When the completed fork leg is mounted on the bike, stop to think about how much fun you're going to have when you're out riding next time. Now move on to the other leg. Before you ride, return the air preload (if you have it) to your preferred settings.

After you loosen the top triple clamp bolt but before you release the lower one, be sure to loosen the fork cap a turn or two. Removing the cap when the fork leg is off the bike will be much easier.

Cartridge forks need to have the fork piston rod disconnected from the cap before the preload spacer and spring can be removed. Take note of the order of the washers and spacers for reassembly. Laying the parts down in order on a clean rag is the easiest method for maintaining the proper sequence.

PROJECT 17 — Replacing Fork Seals

Time: 1 to 2 hours

Tools: Wrenches, sockets, screwdrivers, Allen keys, jeweler's screwdrivers, jack or bike lift, oil recycling container, plastic wrap, PVC pipe or seal driver set, zip tie

Talent: ★★★

Tab: $

Parts: Fork seals, fork oil

Tip: Replace both seals, even if only one leaks

Benefit: Airtight (or is that oil-tight?) forks that don't drip crud all over your brakes

Fork oil seals lead a hard life. Just riding down the road exerts extreme pressure on the front suspension. Add to that potholes or railroad crossings or sudden dips in the road, and you can see what a high-stress job they have. Sometimes they wear out and begin to weep. Perhaps, you noticed a teardrop of oil on the top of the slider during your preride check. You'll need to replace the seal and should minimize your riding until you do. If you're not careful, the oil could drip onto the caliper or disc and render that brake useless.

Even if you only need to replace one seal, replace them in pairs to keep them on the same schedule. The parts are only about $13 for the pair, so don't cheap out.

Owners of traditional forks should follow the directions in Project 16 until you have the fork legs off the bike and the springs removed. Some inverted forks require complete disassembly, while others will slip apart once the cap and the cartridge have been separated. If you have inverted forks, begin with the cartridge removal described in Project 16, then if the inner tube slides out, you'll need to pry out the oil seal with a large screwdriver or tire iron. If the inner tube doesn't slide out, you'll need to pull the fork apart like you would a conventional unit.

Those with traditional forks now stand at a fork in the road (sorry about the pun). To the left is the traditional means of replacing the fork seal: drain the fork, remove the damping rod or cartridge bolt from the bottom of the slider, remove the damping rod/cartridge, remove the stanchion wiper, pry out the retaining ring, and (returning to your prehistoric roots) muscle the stanchion out of the slider. The advantage of this method is that you can check the fork bushing for signs of wear. So, if you suspect bushing wear, follow this path. The disadvantage is that lots of extra steps and sweat are involved. (I've always felt that if I'm sweating while working on a bike, I'm doing something wrong.)

On the other side is the road less traveled—and my personal favorite: all you need are some extremely cheap, fresh motor oil; a catch pan; a jack; a small piece of wood; and a car or truck. Once the wiper and retaining ring have been removed from the slider, fully extend the stanchion and completely fill the fork with motor oil. If possible, make sure there is no air in the system. Reinstall the fork cap. You now have a closed system with nowhere for the oil to go. Lay the fork on top of the catch pan with one end against your garage-door frame. If the cap has an air preload adjuster, drill a hole in the small piece of wood to keep the adjuster from being damaged as you press out the oil seal. Now park your car with its front wheel parallel to a door frame. Place a board across the car wheel and wedge your jack between the fork and the board. Slowly extend the jack. With nowhere for the oil to go, the pressure will push the fork seal out. As soon as the seal slides out far enough that you can pry it the rest of the way with a screwdriver, stop compressing the fork, or things could get messy. See, four-wheeled vehicles *are* good for something!

Remove the fork cap and drain the oil into a recycling container. Like you did when changing the fork oil (see Project 16), pump the fork several times and drain again. Before you slide the old seal off the stanchion, note its orientation. While most fork seals look similar, their orientation can vary from model to model of motorcycle. Closely inspect the stanchion for any dings from stones. Minor ones can be cleaned up with a gentle rub of 400-grit wet/dry followed by 600-grit sandpaper. Use a little WD-40 as lubricant and wrap a rag around the top of the slider to keep any grit out of the fork. Wash the stanchion with

A jeweler's screwdriver is a good tool for getting under the dust seal. Be careful not to scratch the stanchion or damage the wiper.

The oil seal retaining ring is pretty easy to get at with a jeweler's screwdriver or pick.

contact cleaner and a rag. If you find a major ding, take the fork to your local bike shop to have a pro look at it.

Moisten the inner surface of the new seal with fresh fork oil. Carefully slip it over the top of the stanchion and slide it down to the slider. Some mechanics will protect the seal from damage by placing a piece of plastic wrap (from your kitchen) over the top of the stanchion. If you have a fancy seal driver set, simply drive the seal into the slider. If you're cheap (like me), take the old seal, cut out the inner surface, and place it upside down over the new seal. If you're lucky, you can find a piece of PVC pipe that matches the outer diameter of the fork seal perfectly. If not, take a hacksaw and cut out about six sections, evenly spaced around the PVC. Clean up all the grit and place it over the stanchion. Wrap a beefy zip tie around the pipe and tighten it until the PVC fingers exactly match the diameter of the fork seal. Now, gently tap the top of the PVC until the fork seal is completely seated. Remove the old seal and verify that the new seal is deep enough to allow the retaining ring to snap into place in its groove.

After installing the retaining ring, slip the dust seal back over the stanchion. Add fresh fork oil to factory spec, set the oil height, and move on to the next fork leg. Look, no more tears!

Once the truck was in the right position, pressing out the seal took less than five minutes, saving several steps in the process.

SUSPENSION

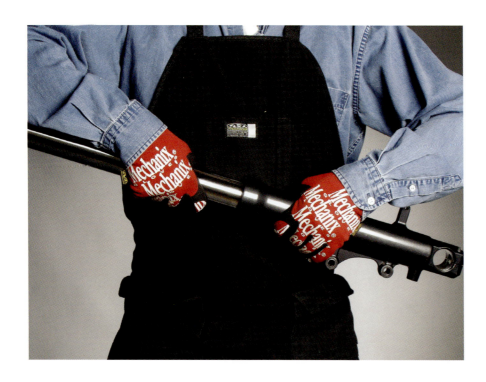

If you have emotional issues to work out, you can remove the stanchion by pressing it in and jerking it out (which sure beats punching your boss). Repeat until it pops free. If this method doesn't work, mount the slider in a vise and put your weight into tugging the stanchion.

Tap the seal into place. You don't need to brutalize it—be gentle. This seal driver may not be pretty, but it costs $3 for 10 feet of PVC. Actual seal driving tools cost significantly more.

PROJECT 18 — Installing Cartridge Emulators

Time: 1 to 2 hours

Tools: Wrenches, sockets, screwdrivers, Allen keys, drill or drill press, rotary tool or deburring tool, jack or front stand, hacksaw, tape measure, impact wrench (optional), recycling container

Talent: ★★★

Tab: $$

Parts: Race Tech Cartridge Emulators, fork oil, Loctite

Tip: Cartridge Emulators only affect compression damping. Adjust rebound damping with the oil viscosity.

Benefit: An impressively well-handling motorcycle

Complementary Modifications: Add aftermarket fork springs (see Project 15)

Cartridge forks have been all the rage in sportbikes for a while. Why should cruiser riders care about this? Well, by allowing adjustable damping rates for compression and rebound, cartridge forks give riders more control in their search for the ideal suspension setup. Folks with damping-rod forks could usually only adjust the spring preload, change the spring rate, or vary oil height and viscosity and hope things went the way they wanted. If the suspension was firm enough, the ride became harsh; if the right level of plushness was found, the bike would wallow, beginning to feel like a mid-1970s Cadillac. Paul Thede, mechanical engineer and president of Race Tech, Inc., recognized the problems inherent in damping-rod forks and designed the Gold Valve Cartridge Emulator to give damping-rod units the plush yet firm ride offered by cartridge forks. Read on for a quick explanation of Thede's invention.

A fork's actions are affected by three forces: friction, spring resistance, and damping. The static friction between the stanchion (inner fork tube) and slider remains constant and is not affected by other suspension changes. Springs know only one thing: position—how much the spring is compressed. The more compression, the more force with which the spring pushes back. Damping responds to variations in road surface in a way that is different from, but complementary to, a spring's response. Since small bumps produce different forces on the front wheel than big bumps, damping responds to the velocity of the fork's compression or rebound.

Traditional damping rods control the velocity of a fork's travel by pumping oil through holes in the damping rod. Since these holes are a fixed size, a compromise needs to be made to accommodate the slower fork movement of rounder

road irregularities and the fast movement of square-edged bumps. As with any compromise, neither situation's requirements are completely met. What's needed is a way to make the fork firmer during low-speed compression and softer during high-speed compression.

By separating the low-speed and high-speed damping portions of a fork's duties, cartridge forks address the compromise inherent in the less sophisticated damping rods. Using shims to restrict the oil flow when the slider is compressing slowly, the bike has a firm ride that doesn't feel mushy and is less prone to squatting or wallowing in corners. When the tire encounters a large or sharp-edged bump, the shims flex to allow larger volumes of oil to move by quickly, giving a plush ride that doesn't hammer the rider's kidneys.

For years cruiser riders could only watch from the sidelines as this suspension advance reverberated through the more sport-oriented motorcycles. In recent model years, cartridge forks have been making their way into select cruiser

SUSPENSION

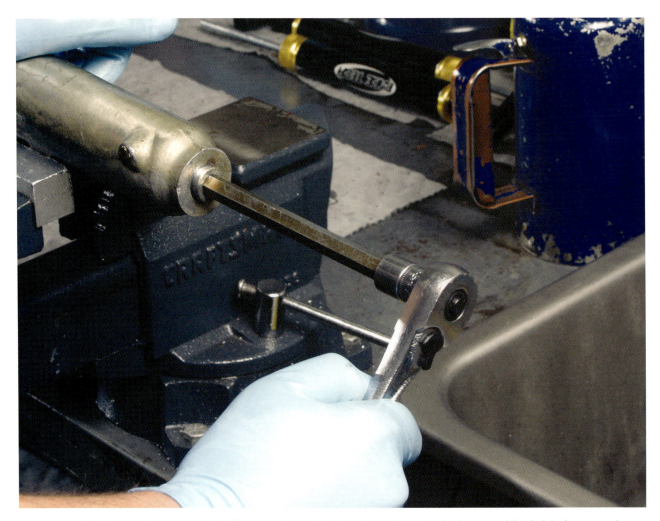

Those who don't have an impact wrench will want to loosen the damping-rod bolt on the bottom of the fork before removing the fork cap.

models (although often in only one fork leg). Thede's deceptively simple invention, the Gold Valve Cartridge Emulator, gives a conventional damping-rod fork a cartridge-like ride by providing tunable compression damping with a reasonable application of elbow grease. Measuring only 14 mm in thickness, the Cartridge Emulator sits atop the damping rod and conducts its business through a flapper valve (called the Emulator plate) and a spring. The Emulator plate has a small hole to control the low-speed compression damping, and the high-speed compression damping is tuned by varying the Emulator spring's preload. Crank in more preload with an Allen key, and more pressure is required to lift the Emulator plate, resulting in a firmer ride. With the compression damping control separate from the damping rod, the rebound damping can be controlled through oil viscosity. Simple, huh?

To install the emulators, support your bike with the front end off the ground. Remove the calipers and use a piece of wire or zip ties to hang them from the bike, taking

Be careful when removing the fork cap. The more spring preload it's under, the more forcefully it'll come out to greet you.

A drill press will simplify the task of opening up the holes in the damping rod, but if you have a steady grip, don't hesitate to use a hand drill. Drill all the way through both sides of the damping rod to speed the process.

the strain off the brake lines. Remove the wheel, fender, and fork brace. Using a wrench, loosen screw-on-type fork caps a turn. Fork caps secured with circlips will be taken care of after removing the fork legs. Loosen the top triple-clamp pinch bolt. Next, be sure to hold on to the fork leg as you loosen the lower pinch bolt. The stanchion should slide out of the triple clamp.

Those without impact drivers should invert the fork and place an Allen key into the bolt securing the damping rod on the bottom of the fork. Rap on the Allen key two or three times to shock the threads. Now, loosen the bolt a bit. Don't remove it yet.

Although not absolutely necessary, a vise to hold the fork tube will make most of the remaining steps easier, but be sure not to over-tighten the vise, which can distort the tube. Use soft jaws or a couple pieces of wood between the tube and the vise. Remove the fork cap by either unscrewing it, or pushing it down and removing the circlip. Lift out the spring. Drain the fork oil by inverting the fork over a suitable container and pumping the tubes until no more oil comes out. Remove the Allen bolt holding the damping rod in the bottom of the lower fork tube. If the damping rod spins inside the fork, preventing its removal, try the time-honored method of inserting a wooden broom handle or dowel to hold the damper rod in place. An impact driver may help too. Once the bolt is removed, the damper rod should fall out of the inverted fork. Don't lose the brass washer under the damping-rod bolt. Often it will stick to the fork but fall out later when you're not looking.

Before proceeding any further, check to see that the Cartridge Emulator fits on the large end of the damping rod, completely covering the opening. Make sure the inner diameter of the fork spring is at least 4 mm larger than the Emulator plate. If your fork uses a "flat top style" damping rod, you will need a special adapter (which should be included with the Cartridge Emulator kit, with rare exceptions) to hold the Emulator in position.

The Cartridge Emulator requires an unrestricted oil flow in order to work its magic. To meet the necessary minimum flow-through, the damping rods need at least six (three sets of two) 5/16-inch compression holes. Since most fork's damping rods are more restrictive than the Cartridge Emulator, the existing compression holes need to be enlarged to the correct size and extra holes drilled, if necessary. Make sure the new holes are drilled perpendicular to, and at least 10 mm center to center from, the adjacent holes to keep from weakening the damping rod. Mark the damping rod and make an indentation with a punch to keep the bit from walking as you start to drill. Continue drilling through both sides of the damping rod. Chamfer and deburr the new holes, leaving a smooth surface on both the inside and outside of the rod. Do not enlarge or make any modifications to the rebound holes.

Start reassembling the fork by sliding the thoroughly cleaned damping rod back into the inner fork tube. Install and torque the Allen bolt in the bottom of the fork tube. Again, the old wooden dowel trick will help hold the rod in place. If the Cartridge Emulator is being installed in a fork with a preload spacer, the spacer will need to be shortened 12 to 14 mm, depending on the model of the Emulator, to maintain the correct preload.

Before placing the Emulator into the stanchion, set the preload on the valve spring. Loosen the jam nut on the bottom of the Emulator, and using an Allen key, loosen the bolt until the spring is completely free. Tighten the bolt until it just touches the spring. Now, count in the turns of preload you put on the spring. Two full turns is the standard starting point for street riders, while four turns should be used for very heavy loads (such as two-up touring). Apply Blue Loctite to the threads before tightening down the jam nut.

Pour the factory-recommended volume of fork oil into the stanchion and pump until no more bubbles appear in the oil. (If you're unsure what viscosity to use, Race Tech's

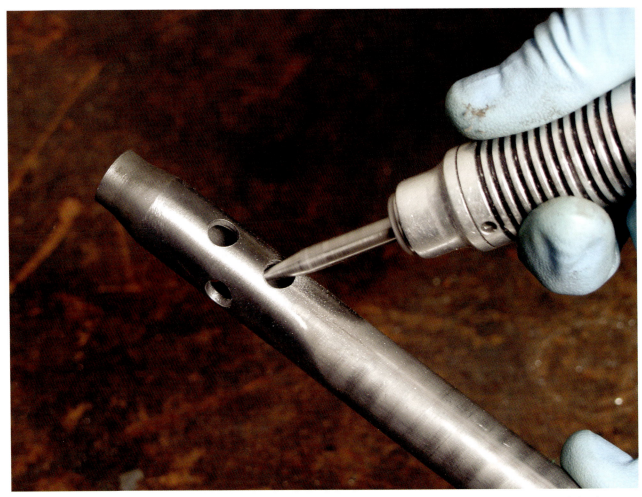

Deburr both the inside and outside of the new holes in the damping rod. You don't want to risk having any metal shavings get into the fork oil and damage sensitive components.

instructions feature a handy table of suggestions.) Install the Emulator with the spring up. Set the oil level with the Emulator installed and the fork completely compressed. Do not set oil level with the fork spring installed (see Project 16).

Replace the fork cap. Pump the fork a few times to check for binding caused by an improperly seated Emulator. Reinstall the fork on your bike and go have some fun. Ride through some corners with an undulating surface, try hitting some bumps, and feel the difference a couple hours just made.

On the left, the new holes will allow free flow of oil to the Emulator. On the right, a stock damping rod shows how big a change we made.

Start with the recommended preload on the Cartridge Emulator. Install them, and make changes to the preload depending on how you want to modify the performance.

The lay of the land:
1. Fork spring
2. Cartridge emulator
3. Damping rod
4. Top out spring

SUSPENSION

Adjusting Cartridge Emulator Damping

Although Cartridge Emulators mimic the function of cartridge forks, they can't match the ease of adjustment that cartridges offer. To adjust a Cartridge Emulator, you will need to remove it from the fork. Don't sweat it—the more you do this, the easier and quicker it gets.

When you have the cap off and the spring removed, lift the Emulator out of the fork. While magnetic tools can accomplish this, they tend to grab the stanchion as you try to lift out the Emulator. A claw tool works much better. Clean and dry the Emulator to make it easier to handle. Loosen the locknut on the base of the Emulator. Now, adjust the Emulator's preload in one-quarter to one-half turn increments, depending on how big an adjustment you feel you need. As you get closer to your final setting, make the adjustments in quarter turns. If you want more compression damping, add to the preload, and for less compression damping, lessen the preload.

Adjusting the rebound damping is an entirely different ball of wax, but pays off big when you take the time. Remember how you only drilled the compression orifices in the damping rod when you installed the Emulator? Well, the rebound circuit now operates separately from the compression. Since the holes will stay the same size in the damping rod, varying the viscosity of the oil will change the rebound damping. Of course, you will need to dump the current oil to change the viscosity. If you want more rebound damping (to minimize the boingy feeling as the fork rapidly rebounds from compression), put in heavier oil, and vice versa. If you discover that neither 10W nor 20W will work and no 15W is around, mix the two oils 50-50 to approximate 15W. You can further tune the rebound by varying the ratios. Just remember to set the oil height correctly after each change.

Although time-consuming, these adjustments are well worth the effort.

PROJECT 19

Installing an Aftermarket Shock

Time: 1 hour

Tools: Wrenches, sockets, torque wrench, socket extenders and/or universal joint, bike jack or lift, zip tie, an optional assistant

Talent: ★ to ★★★ (depending on shock location)

Tab: $$$ to $$$$

Parts: Aftermarket shock

Tip: Mounting a single shock inside the chassis requires more finesse than dual shocks.

Benefit: Better control of your bike's rear suspension, better grip, better handling

Complementary Modification: Lube suspension linkage

Stock cruiser shocks haven't kept up with the march of suspension technology. For the way that many cruisers are ridden, this hasn't really made a difference. Still, many riders prefer the styling of aftermarket units. Other riders, thanks to their riding ability, can sense the limitations of the OE units. Whatever your reason for swapping shocks, you'll generally get a sexier looking, better performing bike with the application of an aftermarket suspender or two. Even if you're not adding an aftermarket piece, this project will be helpful to those who need to remove their shock to send it off for revalving, a simple rebuild, or installation of a different rate spring.

Begin by putting your bike on a jack or bike lift. Next, you'll want to secure the front end so that it doesn't roll away while you're elbow deep in the bowels of your bike. While some people perform this modification with the front wheel snugged up against a wall, I've found that a zip tie around the front brake lever works just fine for a simple shock swap.

If your bike will be sitting without a shock while it's out for freshening, you should consider finding a way to secure your bike. Regardless of how long you expect the project to take, don't lie under your bike until you're

Lift your bike so that the shock fully extends, but the rear tire doesn't leave the floor.

SUSPENSION

Space is really tight, but with the rear wheel supported, the shock should slip right into position. You may have to adjust the bike lift up or down slightly to get the eyelets on the shock to line up with the chassis.

certain that it is safely supported. Remove any parts (like the seat, saddlebags, side covers, or the exhaust system) that block unfettered access to the area around the linkage. During this project, you'll be adjusting the height of the bike on the jack or lift to allow the eyelets on the shock to line up with their mates on the bike.

Dual-shock bikes are much easier to work with since you'll always have at least one shock on your bike (that is, if you're maintaining stock ride height). You'll still need to jack the bike up to take the load off the shocks and avoid having the chassis settle when you remove one. Other than that, you're just unbolting one and torquing on another. Easy peasy.

If your cruiser only has one shock, you'll be working in the space inside or under the frame. A single shock that acts directly on the swingarm is the second easiest to replace. While space will be tight, you'll only have to remove the two bolts affixing the shock to the chassis. Still, the torque values are reasonably high on the bolts. If you're working alone, be prepared to exert yourself a little as you break the locknut and bolt free. If you have an assistant, you should work on opposite sides of the same bolt. If you can't slip the bolts free when you have them loose, try raising or lowering the jack supporting the bike until you've freed up the tension on the bolts. Before you yank out the bolt, though, check to see how easily it slides free. Try adjusting the height of the jack until the bolt just slips out. You'll have a much easier time with reassembly if you take the time to do this. A stock length shock should slip right back into place, while a shorter shock will require that you lower the bike on the jack until the mounts match up. Attaching the shock to the top mount and bringing the lower mount up to the shock is the easiest method here.

Cruisers with the shock attached to some sort of linkage require that you partially disassemble the linkage to get to the shock. Begin with the base of the shock. The same rules concerning the shock bolts in the previous paragraph apply here, too. Look closely at the bolts securing the tie rods (or dog bones). On many bikes, you will have difficulty removing the bolts for either the top or bottom end of the rods. Determine which bolt is easiest to remove and lower the linkage out of the way. Remove the bolt securing the top of the shock to its mount and carefully lower the old shock out past the swingarm.

Carefully place the new shock into position. Some shocks are easier to slip into place from above the swingarm rather than from below. Place the top mounting bolt through the shock's eyelet and let it hang in place. If your new shock has a remote reservoir, you need to find a place to mount it. If the shock's instructions don't have a recommended

Many aftermarket shocks will have rebound damping adjusters that allow you to modify the shock's behavior as it decompresses after absorbing a bump.

position, find a place that you can securely mount the reservoir so that it does not interfere with the rear wheel travel or with the rider's leg or foot. Using the supplied rubber spacers and hose clamps, loosely mount the reservoir and check that the shock's braided stainless-steel line doesn't abrade the frame or any other part of the bike. You can buy line covers that wrap around the line like a spring at your local bike shop. Once you are certain about the reservoir position, tighten the hose clamps.

Reassemble the suspension linkage in the reverse order of your disassembly. A light coat of grease on the bolts' shafts (while keeping the threads clean) will help them slip into place and prevent corrosion. You may find that you have to rotate the shock shaft slightly to help the clevis slide over the linkage. Again, if you have trouble lining up the bolt holes, adjust the length of the bottle jack supporting the rear of the bike. Proper torquing of the bolts is essential for keeping everything where it belongs.

Once you have the bike buttoned up and the jack removed from the rear wheel, ignore the irresistible urge to jump on the bike, go for a ride, and then twiddle with the damping adjusters that probably came with your fancy new shock. First, you need to set the sag and damping to the shock manufacturer's recommendations (see Project 14). Once you have this last step completed, then go play with your new toy.

Dual shocks are almost too easy to replace. Simply unbolt one and install the other. You still need to support your bike with a jack, though.

PROJECT 20 | Lowering a Cruiser

Time: 3 to 5 hours

Tools: Wrenches, sockets, torque wrench, socket extenders and/or universal joint, pick, preload adjustment tool or long screwdriver and hammer, dead-blow hammer, bike jack or lift, metric tape measure, impact wrench (optional), cartridge fork tool

Talent: ★★ to ★★★★

Tab: $ to $$

Parts: Lowering blocks, shortened shock or linkage

Tip: These modifications limit suspension effectiveness and ground clearance for normal riding.

Benefit: More oohs and ahhs from observers, less shifting of weight on aggressive drag-strip-style launches

If straight-line blasts do more for you than twisty roads, you are probably considering lowering your bike. Or maybe you think that cruisers only look their best when they are slammed down to the ground. Either way, the style has its roots in drag racing. The one goal of drag racing is to get all the horsepower to the ground without looping the bike. To that end, lowering a bike for the strip is a time-honored means to achieving this goal. With the center of gravity (CG) closer to the ground, the weight shift under acceleration is minimized, as is the leverage that could lift the front wheel off the ground.

Front Suspension
Since you can't just drop the triple clamp down on the fork without running into clearance problems between the tire or fender and the chassis—not to mention the ridiculous look of having the fork legs poking way out of the top of the triple clamp—you've got two choices for lowering the front end: temporary external straps and internal lowering blocks. While some sportbike riders resort to tie-down straps as a temporary way of dropping the front for street races or Friday Night Grudge Matches at the local drag strip, cruiser riders would never resort to such a butt-ugly means of lowering their bikes—even if it was for money. So the only real method for lowering a cruiser (and, hence, the only one covered here) is the internal mechanism that keeps the fork from fully extending.

Damping-Rod Forks
To drop the front of a bike with a lowering kit, you need to disassemble the fork. The block provides a mechanical stop that keeps the fork from fully extending. For damping-rod forks, a spacer is inserted between the stanchion and the slider, keeping the stanchion from being able to extend fully out of the slider. Installing a lowering kit in a damping-rod fork is no more complicated than changing an oil seal as described in Project 16 (although you will have to use the pull-the-bushing-free method). Once you have the stanchion out of the slider, simply remove the bushing and slip the lowering block into position. Since you've got the fork apart, why not install a fresh bushing? Coat the inner surface of the bushing with fresh fork oil before slipping it onto the stanchion. Place the stanchion in the slider all the way to the bottom of its travel. Use a seal driver or your homebuilt PVC version to seat the bushing as described in Project 16. Follow the reassembly instructions from the remainder of that project.

Cartridge Forks
In cartridge forks, the spacer is inserted into the cartridge and requires considerably more mechanical experience. If you're adventurous and trying this for the first time, modify one fork leg at a time so that you have an assembled, functional one for comparison.

Upgrading your cartridge fork's valving does not involve completely disassembling the fork, but you do completely disassemble the delicate parts of the cartridge. If you're uncomfortable with small parts, meticulous cleaning, and assembling delicate items, perhaps you should ship off your fork to your favorite suspension company for the job. Those adventuresome souls who prefer to perform the work themselves should proceed with a vigilant eye for detail.

If you decide to do it yourself, remove the fork from your bike and mount it in a vise. Owners of traditional cartridge forks should disassemble the cap and damping piston rod as described in Project 15. Some inverted forks (such as those on the Yamaha Warrior that are based on the R1 sportbike fork) require additional steps and tools, so consult your factory service manual. If you can't find the bolt to release your cruiser's cartridge from the fork cap, your bike requires a cartridge fork compression tool (available as a factory service part from your bike's manufacturer, or from an aftermarket company such as Race Tech) to compress the fork spring to reveal the locknut securing the cartridge to the cap. Once the cartridge is compressed enough to reveal the nut, use a pair of wrenches, loosen the locknut and spin the cap off the cartridge piston rod. Put the fork oil in a container for recycling. All of the cartridge disassembly, valving, and shim information applies to both styles of fork.

After the fork cap is off either style cartridge fork and the oil has been removed, free the cartridge itself from the fork. To accomplish this, you need another cartridge tool to hold the fork internals in place while the bolt securing it to the bottom of the assembly is removed. Although sometimes you don't need this tool to disassemble the fork, don't cheap out or you may find yourself with a partially assembled fork, after business hours, on a holiday weekend. After the bolt is removed, the cartridge should simply lift out of the fork by the piston rod. Completely clean and dry the cartridge before attempting to disassemble it.

The compression valve is located on the bottom of the cartridge. The valve on some models screws to the bottom of the cartridge body, usually requiring only a pair of wrenches and elbow grease to remove it. Others have punch marks 15 mm from the bottom of the assembly. These marks will need to be drilled out with a 3/16-inch drill bit. Do not drill any deeper than is required to penetrate the cartridge tube. A piece of electrical tape wrapped around the bit can tell you when you've drilled far enough. Once the holes have been drilled, push the compression valve holder into the cartridge body approximately 5 mm to gain access to the wire clip securing the valve. Use a small screwdriver or pick to remove the clip, and the valve should slide free. Be sure to deburr the holes on the interior of the cartridge tube so that the O-ring doesn't get damaged after reassembly.

Screw-on-type compression valves can be downright ornery at times, thanks to the thread lock used on them. If you have trouble with one of these, try a couple of taps with a hammer on the tube directly outside the threads (not above, which could damage the tube). More stubborn valve assemblies might even require that the cartridge tube be heated with a blow torch.

Once you have the valve separate from the cartridge body, you will need to remove the locknut from the piston rod and slide it free of the cap. Slip the lowering block over the piston rod. Before you reassemble the cartridge, make sure that all the parts are clean and free of grit. When you reassemble the fork, use fresh fork oil. You will also need to shorten the preload spacer the same length as the lowering

The locknut securing the fork cap to the cartridge piston rod is hiding behind the preload spacer. You'll need to compress the spring with a special tool to gain access to the nut.

block. If you don't, you won't be able to get the fork cap back on. Depending on the type of kit you installed, your fork should now be 2 to 3 inches shorter than stock.

Rear Suspension

Lowering the back of your bike is a simple operation. Most aftermarket companies offer shorter shocks for the purpose of lowering cruisers. Simply install a new rear shock as described in Project 19. Yes, this sounds like a vast over-simplification when compared to the previous detail on the fork. However, it is essentially true. A few aftermarket companies will modify the swingarm on select models to lower the back end (usually in conjunction with enlarging the swingarm to accommodate 250 or wider tires). Depending on the linkage construction, you could alter the length of the tie rods (aka, dog bones), but this is exceptionally rare in cruisers and requires custom building of parts since none are available, that I know of.

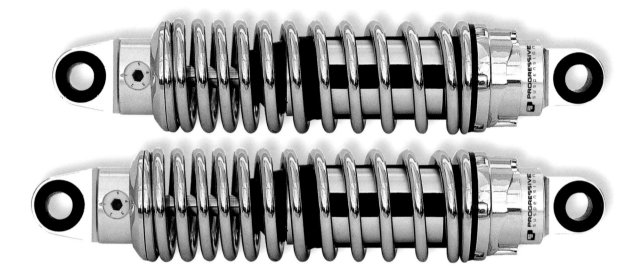

These two shocks are for the same bike. The shorter one simply lowers the bike. *Photo courtesy of Progressive Suspension*

Lowering Cautions

Anyone who's looking to lower their bike needs to be aware of the compromises involved. First and most important, you will decrease your cornering clearance. To put it in plain English, you can no longer lean your bike over as far as its designers intended. If you even only occasionally scraped your foot pegs, floorboards, or any other bike part (such as the exhaust) while riding, you should avoid lowering your bike. You're already approaching your cruiser's cornering limitations, as it is.

If you're still hot to make this modification, there are other considerations to be aware of. For example, you will need to shorten the sidestand. You need to make sure the stand allows the bike to lean over far enough to remain stable and not fall over. Finally, if you do lower your bike, carefully build up to your new maximum lean. Flicking your bike into a corner before you know where the hard parts will drag could have you touching down hard enough to lever a wheel off the ground—and then you'll *really* touch down.

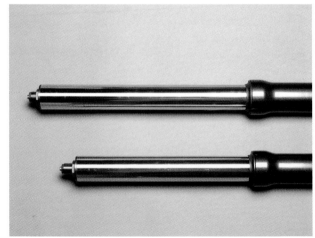

These fork legs are identical except for the lowering block in the shorter one.

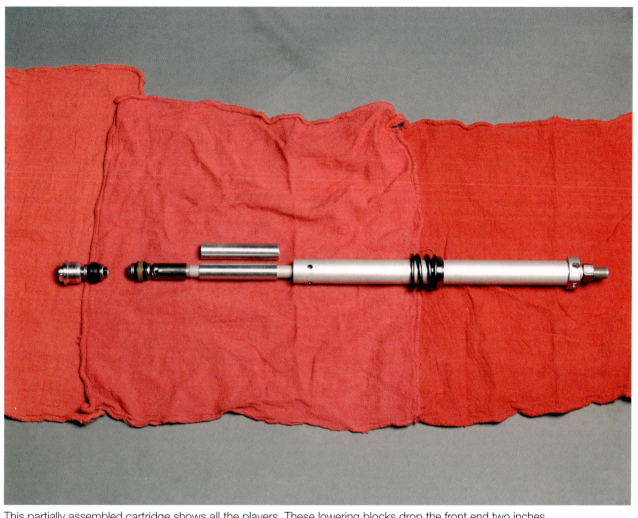
This partially assembled cartridge shows all the players. These lowering blocks drop the front end two inches.

SECTION 5
ENGINE

Projects 21–30

PROJECT 21	Installing a Full Exhaust System

Time: 1-2 hours

Tools: Wrenches, sockets, universal joint sockets or universal joint socket adapter, 3- to 4-inch extension, Allen keys or Allen sockets, torque wrench, rubber mallet or dead-blow hammer, spring puller, flashlight, high-temperature grease, WD-40, soft cloth or work matt

Talent: ★★

Tab: $$$$

Parts: Aftermarket exhaust system, exhaust manifold gaskets

Tip: Tighten fasteners from front to back to ensure you aren't pinching the system cock-eyed against the chassis.

Benefit: Maximize engine performance

Complementary Modifications: Install jet kit (see Project 23) or retune EFI (see Project 25) for optimal power

No matter what motorcycling market you look at, one thing remains the same: motorcyclists love aftermarket pipes. Metric cruiser owners are no different. A recent survey found that one-fourth of the people polled said they planned to buy an exhaust system for their bike in the next year. If you apply this percentage across the entire cruiser market, you begin to understand how a representative of an aftermarket company that makes high-quality accessories for cruisers could tell me that, even with all their billet products, the single largest portion of their business still comes from pipes.

Weight and back pressure (and, by extension, lower peak horsepower) have traditionally been the shortcomings of the EPA noise-restricted stock systems. Aftermarket systems usually offer the benefit of lighter weight and more horsepower in portions of the rpm range, although some carry the penalty of excessive noise. One thing that many people don't know is that many aftermarket systems have flat spots that are much larger than those on the stock system.

Aftermarket exhaust systems fall into two categories: full systems and slip-ons. The full systems replace the entire stock system. Some headers have various tapers and crossovers to enhance low- and midrange torque while still improving top-end power, although this is less prevalent on V-twins than multis. Most have heat shields that serve to protect the rider from the scorching headers and to put enough air between the exhaust gasses and the chrome plating to keep it from bluing. On some bikes, the heat shields are clamped on while others are an integral part of the header.

How do exhaust systems work their magic? The common generalization of internal combustion engines is that they are really nothing more than big air pumps. Like most generalizations, this one has a basis in truth while still being a vast exaggeration/oversimplification, but it serves as a good starting point. If we were only looking for peak horsepower, we would want to remove as much of the restriction inherent in an exhaust system as possible. This way, the engine would be unhindered as it pumped through the maximum volume of atmosphere of which it was capable. While drag racers may live for peak power in a wide-open throttle environment, most motorcycles spend little time at full throttle. Instead, the throttle is almost constantly being adjusted. Not surprisingly, the engine runs over a variety of rpm, too.

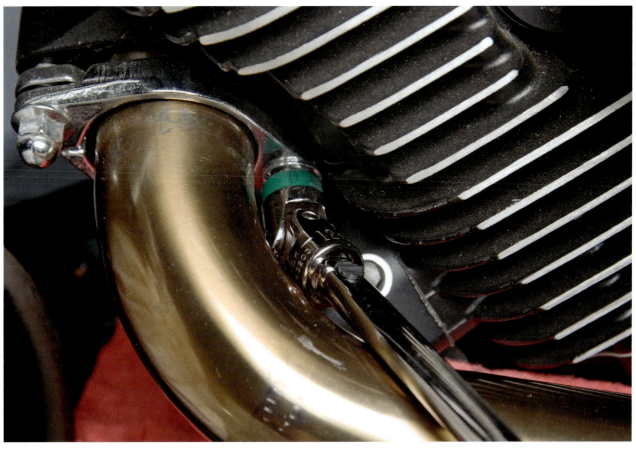

Space can be tight behind the radiator or the rear cylinder. Accessing the manifold nut can be just about impossible if not for a universal joint socket.

Straight pipes that allow maximum peak horsepower will suffer at other rpms. So, where's the advantage? Here is where the pump analogy falls apart. The individual pulses of expanding gasses flow out of the engine in waves. When they hit expansions or restrictions in the exhaust system, the waves reflect back in both beneficial and harmful ways. You see, the exhaust system can also act as a pump to help draw the spent charge out and the new charge into the cylinder—if the pulses are in synch with the valve openings. However, if the pulses are working against the flow by either preventing the cylinder from filling or causing the air to pass through the carburetors more than once, a process called multiple-carburetion (not a factor for fuel-injected bikes) makes the charge burn inefficiently. Either way, the result is a flat spot in the power curve. (For more information on this, read Kevin Cameron's excellent *Sportbike Performance Handbook*.)

So long story short, the design of a pipe will make it operate better at some rpms than others. This is where the exhaust pipe comparisons published in motorcycle magazines (such as those of my former employer, *Motorcycle Cruiser*) can be useful. If you look at the dyno charts, you'll

Screw both bolts in finger tight. You want the bolts to pull in evenly on both sides of the manifold. Before reaching for your ratchet, make sure the entire system is installed and in proper position.

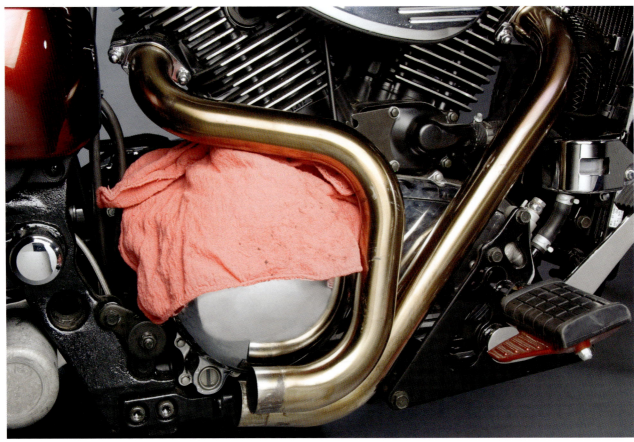

Working your way front to back, install the headers into the muffler. You might not believe it, but it can take some finesse to get everything lined up.

see where the power is better—or worse—than stock. With an honest assessment of your riding style, you can choose the pipe that gives you power where you need it the most. For most of us, fattening up the midrange and extending the peak horsepower output is ideal. Riders in need of ego augmentation can simply look for the biggest peak numbers.

Finally, no preamble about installing aftermarket pipes would be complete without some comments about noise. First, the use of loud pipes probably has the biggest negative impact on motorcyclists in the eyes of the non-riding public than any other motorcycle activity. While that straight pipe may be music to your ears, and that extra three horsepower impresses your riding buddies, you are also spreading ill will in every direction. And the junk about loud pipes saving lives is just a bunch of arrogant, self-delusional hoo-ha that is not supported in accident studies. Use common sense and consider street-reasonable baffles instead of wide-open units. Of course, if you're shopping for a pipe based solely on how it looks and that it gives your bike some "rumble," just go with what attracts you.

The physics of exhaust pulses are such that the only thing that can quietly flow a healthy volume of gasses is a large muffler. This puts big twins at a disadvantage. While a 1300-cc V-four only pumps out the contents of a 325-cc cylinder on each pulse, a 2000-cc twin's pulse is three times as large and there-

Don't forget to wipe down the system completely before you fire it up for the first time. Otherwise, you may get to look at your fingerprints every time you go for a ride.

fore more difficult to quiet without reducing power output. Consequently, the best performing exhaust systems for high-displacement twins will have two high-volume (as in large) mufflers. Just take a gander at the Warrior's stock muffler.

Installing a Full Exhaust System
Begin by placing your bike in first gear on its sidestand. Next, remove any bodywork that will prevent access to the exhaust manifolds. Unbolt the stock muffler from its mounting bracket. If necessary, unbolt the muffler from the collector outlet. You may also find clamps pinching pipe connections. If you have trouble pulling pipe sections apart, try spraying some WD-40 into the seams and letting it soak a bit. Tapping the offending part in the direction you want it to move with a rubber mallet can also help. You may find expansion chambers tucked away under the bike that need some additional mounts loosened before they'll slide free.

Eventually, you'll work your way to the front of the engine. On some bikes you may need to loosen and tilt the radiator and/or oil cooler up out of the way to give you access to the exhaust manifold nuts (or bolts). You'll find it helpful to use a socket with a universal joint to reach the fasteners. Loosen all the manifold nuts before removing them. This will keep the header from falling off while you're unscrewing the rest of the nuts. Once all the nuts are removed, wiggle the header loose from the studs. Make sure to remove all the exhaust manifold gaskets. Sometimes they're hard to see. Use a flashlight to check.

Installation of the new system will be the reverse of the removal process. You'll start at the engine and work your way back. Don't tighten any of the fasteners any more than finger tight until you have the entire system installed and adjusted into its final position. Otherwise, you risk torquing the system in a way that places constant forces on it. This could transfer annoying vibrations to the chassis that may lead to premature wear (metal fatigue) and failure of your expensive new exhaust system.

Before you mount the headers, you'll want to secure the manifold gaskets with a dollop of grease. Otherwise, you'll need three hands to hold the gaskets in place while jockeying the header and its assorted flanges into position. When you start the engine, you probably won't even notice the grease burning off, since new exhaust systems smoke on their first run-in anyway. Screw the nuts or bolts so that they are finger tight evenly across the header tubes. You don't want the clamps cock-eyed or you may develop exhaust leaks.

Depending on your exhaust system, you'll either install the mufflers individually or in pairs. If you mount both mufflers to a bracket before sliding them into position, keep the bolts loose so you have some wiggle room to attach the canisters to the headers. A spritz of WD-40 on the portion of the pipe that slips inside of the other will make the parts slide easier. Again, this will burn off without you noticing. Once you have the canister(s) attached to the header, you may need to do some jiggling or rotating of the components to get all the mounts to line up correctly. Tighten all nuts and bolts finger tight and check the alignment of the system. You don't want the entire exhaust system to have an unnecessary load on it once you tighten the bolts. Install any retaining springs. Beginning with the manifold nuts or bolts and moving rearward, torque the systems fasteners to spec. Install the header heat shields with their clamps rotated in such a way that they will be out of sight but still within reach of a screwdriver or socket.

Slip-On Exhaust
Slip-on exhaust systems deliver much of the look of an aftermarket system and some of the performance of a replacement system at a significantly lower price than a full aftermarket system—particularly for V-fours. Installation is much simpler for novice mechanics, too. Before you begin wrenching, take a look at the fasteners securing the stock components you will be removing. Some bikes may only require a couple fasteners to be loosened, while others will need a little more work. If you have trouble slipping the muffler off the header, a quick tap with a rubber mallet will usually help the OE canister slip free. Once again, stuck components can benefit from some WD-40. Follow the same precautions of installing all mufflers and making sure they line up before securing the bolts beyond finger tight.

With both slip-ons and full systems, before you start the engine, wipe down the entire system to take off your grubby fingerprints. Pay special attention to the visible areas of the pipe. Some manufacturers recommend using rubbing alcohol to cut the oil. Both stainless steel and chrome will permanently display your fingerprints once it heats up to operating temperature. You have been warned.

Tilt the radiator (and oil cooler) back into position and tighten down all the mounting bolts. When reinstalling the bodywork, make sure that there is adequate clearance between the pipe and the plastic. If you find interference problems, double check the pipe for correct mounting.

Careful owners will want to take some time to season their pipes before taking the cruiser for a ride. Start your engine and let it idle for a couple minutes, shut it off, and let the pipe return to ambient temperature. Repeat this process two more times adding a couple of minutes to the run time. Although all single-walled pipes discolor from heat, performing this process will help to minimize the initial bluing of your headers and chrome mufflers. Don't forget that new pipes will smoke for a few minutes, so don't be surprised. Check all the fasteners, particularly the manifold bolts, after about 50 to 100 miles and retorque, if necessary.

Now for the fun part: Get out there and try out your new exhaust and additional street cred. To get maximum performance from your new exhaust system, consider rejetting or altering your fuel injection. Your bike may run fine without it, but it will run better with the proper fuel mixture.

Sometimes just changing the exhaust tip or buying a pipe of a different length will dramatically alter the look of a bike. *Photo courtesy Cobra Engineering*

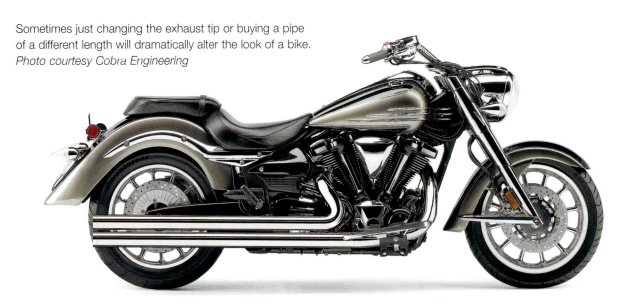

So many cruiser owners buy drag-style pipes that you might forget that there are a multitude of styles out there. This 2-into-1 megaphone looks good, and you won't see it on every other cruiser you pass. *Photo courtesy Cobra Engineering*

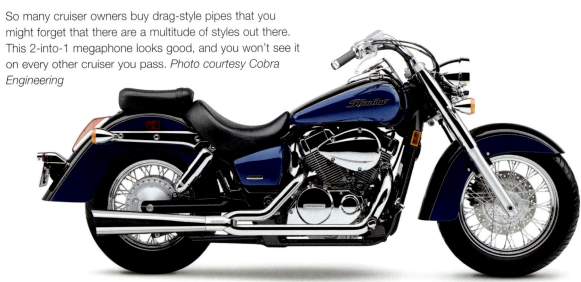

Staggered dualies are another popular style of pipe. Many manufacturers offer a variety of tip styles other than just the slash cut shown here. *Photo courtesy Cobra Engineering*

Aftermarket pipes for some bikes, by virtue of their multiple cylinders, can be pretty pricy. Slip-on systems are a less expensive and just as stylish option. A well made slip-on will make almost as much power as a full system, too. *Photo courtesy Barons Custom Accessories*

Just because pipes usually come in just chrome or stainless steel doesn't mean your custom vision can't incorporate other heat-tolerant finishes. *Photo courtesy Barons Custom Accessories*

PROJECT 22

Synchronizing Carburetors and EFI Throttle Bodies

Time: 1 hour

Tools: Wrenches, sockets, ratchet, #2 Phillips screwdriver, clean rags, carb balancer, box fan, auxiliary fuel tank

Talent: ★★

Tab: $

Parts: None

Tip: Make sure all vacuum leaks are plugged or you will get false readings.

Benefit: Smoother power delivery

Complementary Modifications: Clean/replace air filter (see Project 5)

Like other mechanical devices, carburetor and EFI linkages periodically need to be adjusted to ensure that, when you roll on the throttle, the same thing happens to every throttle body. Since motorcycles generally have as many carburetors as cylinders, multi-cylinder engines use linkages to open the butterfly valves as you roll on the throttle. (Note: This parity of carbs/throttle bodies to cylinders is by no means an absolute. Many bikes such as the Road Star or Vulcan 1500/1600s have only one mixer, thus negating the need for synchronization. Do a little research in your factory service manual.) By making sure that all the butterflies are open the same amount at idle speed, the throttles should allow the same amount of air through at all throttle positions and engine speeds. If carburetors or EFI throttle bodies (hereafter lumped together as carburetors) are out of sync, the cylinders will receive different-sized charges. The symptoms can be as mild as rough idling or an uneven engine sound at partial throttle. Severe cases can cause surging while at any constant throttle setting.

The reward for spending an hour syncing your carbs is a smoother-idling, quicker-revving engine. The only tools required other than typical home mechanic's sockets and wrenches are a vacuum gauge and an auxiliary fuel tank (for bikes that don't have a fuel pump). While you can spend big bucks on a fancy digital vacuum gauge if you want, you only need one that measures pressure by how far it sucks mercury up a tube. Motion Pro sells two reasonably priced versions of their carb tuner through distributors and motorcycle dealerships. A purpose-built auxiliary tank supplies fuel to your engine while keeping any gas from dripping on the engine's hot parts. This is also available from Motion Pro.

Since your bike will be running, make sure your work area for this project has good ventilation. Start by removing

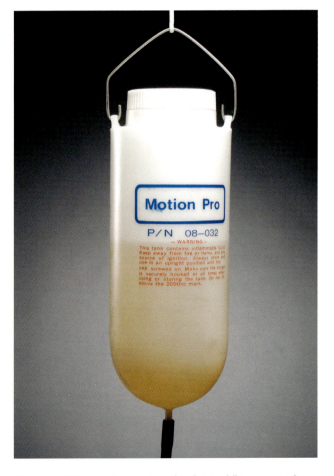

Since you'll be running your engine for a while, you need a supply of gas for your carburetors. Hang the auxiliary tank above the level of the carbs but out of the way for your wrenching.

Your bike will either have a capped nipple or a threaded plug. Some manufacturers make your job easier by routing hoses out from under the throttle bodies. You'll need to either attach your carb balancer to the nipple or hose.

the seat, gas tank, and anything that gets in the way of your access to the carbs. Be careful when disconnecting the fuel line from the petcock, particularly if the engine is hot. Hold a rag under the open line until all the gas has drained out. To avoid damage to the petcock and possible fuel spillage, place the fuel tank on an old tire to keep it from tipping over. Smart mechanics label the various hoses as they are disconnected from the air box and tank.

Now, take a look at the hoses around the carburetors or throttle bodies. If you're lucky, the manufacturer has pre-connected hoses to allow for easy synchronization. Your factory service manual will help you find them. If you don't have those hoses and can't access the nipples on the throttle bodies with the air box in place, disassemble and remove the air box to give unfettered access to the carburetors. (If you haven't done so lately, now would be a good time to make sure the air filter is clean and in good repair.) Hook the auxiliary tank to the fuel line and hang it so that the tank is higher than the carburetors. If your bike uses a fuel pump that is built into the gas tank, you will need to find a way to keep the fuel line, return line, wiring harness, and tank mounted to feed the carbs—a tricky operation.

Locate and uncover the ports into the intake tract. You will find either bolt plugs or capped nipples. Your carb balancer should include threaded adapters to fit the port. While most port threads will be 5 mm, some bikes use 6 mm threads. Once you have the adapters screwed in, attach the hoses to the nipples. Starting at the number 1 cylinder (the front cylinder on V-twins or the front left on other multi-cylinder configurations), connect the first hose and move across the front cylinders (if the bike has them) left to right before the rear cylinder(s) making sure to keep the hoses in order. Find a convenient place to hang the carb balancer so that you can see it while the engine is running.

Even water-cooled bikes need to have air moving around them to maintain proper operating temperature. A box fan will work nicely.

After, and possibly while, adjusting the linkage to improve throttle synchronization, you will need to adjust the idle speed back to factory spec.

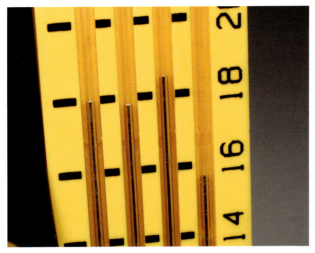
Throttle bodies 3 and 4 are out of sync with each other. The lines on the scale mark off 2-centimeter increments.

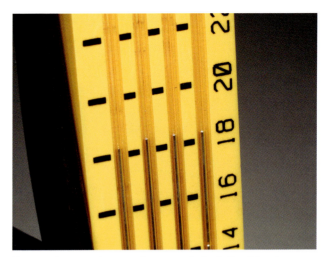
You want your columns of mercury to look like this when you're finished with your adjustments. Be sure to test the synchronization at cruising rpm to make sure that they maintain their relative closeness.

Before you start the engine, make sure that the hoses do not interfere with the throttle linkage. Also, make a thorough visual check to see that all possible vacuum leaks have been sealed. When you start the engine, be careful if you need to give it some gas. Blip the throttle too energetically, and you'll get the pleasure (not) of seeing the mercury sucked out of the carb stick into the engine.

If you value your brain cells, don't forget to make sure you have proper ventilation. While the engine warms up, listen for any vacuum leaks you may have forgotten to seal. After your engine reaches operating temperature, make sure you have a fan blowing across the radiator (or engine in air-cooled configurations) to help keep your bike from overheating. If you let it get too hot, the carb syncing may not be accurate. Finally, make sure the idle speed is set to factory specs.

Most bikes, be they carbureted or fuel injected, synchronize the throttle bodies by adjusting the linkage connecting the butterfly valves that control the flow into th engine. However, some fuel-injected bikes simply require that you adjust an air screw in each throttle body to get it to match the one body that doesn't have an air screw. Check your factory service manual. Either way you adjust the synchronization, the carb balancer displays the same information. The vacuum created in the intake tract will draw the mercury up from the reservoir in the bottom of the carb tool. There will always be some variance between the columns, but most manufacturers say that a one-half- to one-inch difference of mercury level is fine. However, adjusting the columns of mercury so that they are as close to identical as possible is worth the minimal effort it requires. Twins will have only one screw to adjust the synchronization. So, you're done at this point.

Multis, be they inline-four sporty bikes or V-four cruisers, require the process be done in steps—as do EFI systems that use the butterfly valves (rather than air screws) to synchronize the throttle bodies. Begin by finding the adjuster screw between the number 1 and 2 cylinders. Turn it until you have the two carb-balancer readings identical. Blip the throttle slightly and let the engine return to idle. Make an adjustment if necessary. Once you're happy with the results, switch to the adjustment of the number 3 and 4 cylinders. Again, once the two sets of two carburetors are set, find the adjuster screw between the number 2 and 3 cylinders to balance the two pairs of butterfly valves.

Now that the carbs are in sync, raise the rpm to 3,000 and hold it steady. The columns of mercury should settle at a consistent height. If one or more of the columns rises to a lower level than the others, a vacuum leak or some other problem will need to be identified and corrected for your bike to run its best. Common problems are a worn slide that is sticking or something failing in the linkage. Finally, adjust the engine's idle speed back to the factory specs if it changed. Carefully remove the carb tool and reposition all vacuum hoses and plugs. Start the engine again to see if there are air leaks—the idle speed should stay the same. Replace the tank, paying special attention to the fuel and vacuum lines connected to the petcock. Your bike's engine should now idle smoother and rev cleaner.

PROJECT 23 | Installing a Jet Kit

Time: 2 hours

Tools: Wrenches, sockets, ratchet, Phillips screwdriver, flathead screwdriver, drill with screw attachment (optional), Allen keys, float height tool (if required), pliers, drill, rags

Talent: ★★

Tab: $

Parts: Jet kit

Tip: Installing a jet kit is more than just tossing in a new main jet—take your time and work carefully.

Benefit: A bike that carburets cleanly throughout the entire rev range

Complementary Modifications: Clean/replace air filter (see Project 5)

Common knowledge dictates that the first hop-up item most riders purchase is an aftermarket pipe. While most aftermarket exhaust manufacturers say rejetting is not required, they generally recommend installing a jet kit for optimum power. However, experienced mechanics will tell you that only about half of the folks who have fitted freer-flowing exhaust systems to their bikes have also installed jet kits. The people who refrained from installing jet kits usually give one or both of the following statements: "The jet kit was too expensive," and/or "The pipe improved my bike's power so much, I didn't need a jet kit."

When you open the box to your new jet kit, the few brass jets (one per carburetor) and stainless-steel or titanium needles don't look like much for the $60 to $100 you just plopped down, but the CD that expensive computer software rides on doesn't look like much, either. Remember, you're not only paying for the machining of the jet kit's little pieces of metal, but also for the development time. Sometimes, a kit may take several weeks of tweaking to reach perfection. (For more information on what it takes to develop a jet kit, see Project 26.)

All riders benefit from having a well-carbureted bike. While an aftermarket pipe may make a bike more powerful, properly jetting a bike's carbs will make an engine not only produce even more power, but also improve the quality of how the power is delivered—whether an aftermarket pipe is bolted on or not. Many bikes come from the factory with extremely lean jetting to meet EPA requirements, and sometimes something as simple as raising the needle with a shim can produce a night-and-day difference in your bike's performance, particularly off-throttle to on-throttle transitions. When you order a jet kit for your bike, be sure to order the kit based on the current tune of your bike. Don't get a Stage III kit just because you plan to make big changes later on. If you've only got a pipe right now, jet for your bike with the pipe.

Begin by placing the bike in gear on its sidestand or on a bike lift. If your cruiser hides the carburetors under the tank, you'll also need to remove the gas tank, air box, and any other crud that keeps you from gaining direct access to your carburetors. Make reassembly easier by labeling all hoses and wire connections. Owners of V-twins that hang the carburetor off the side of the bike will want to remove the tank to gain easier access to the carb and its associated doodads. Disconnect the throttle cables from the bell crank. (From here on, the use of the word "carburetor" will also apply to multiple carbs.) Loosen all the clamps securing the carburetor mouth to the rubber and the intake manifold. Gently wiggle the carb up and down until it pops free of the boot. Don't put it down just yet. Hold the carb over a suitable container and drain the float bowl. Before you do anything else, place clean rags or paper towels in the intake manifold to keep any nasties out.

Prior to exposing the carb's innards, you might as well remove the brass plug covering the idle screw. Most jet kits will include a drill bit and self-tapping screw to remove the plug. If you don't have one, use a 1/8-inch drill bit and any wood or drywall screw you have sitting around. To keep from drilling too deeply past the plug and possibly damaging the idle jet, wrap a piece of tape around it about 1/8 inch above the tip. Don't drill any deeper than the tape. Insert the screw far enough to have a solid grip on the plug with the threads. Using a pair of pliers (locking pliers if you like), pull the plug free of the carburetor body. Before you go any further, insert

In order to access the idle jet, you need to remove the EPA-mandated cover. Note how the drill bit is taped to prevent damaging the adjuster screw below the cover.

If you don't have a large flathead screwdriver, use an 8-millimeter socket to remove the main jet. Install the new one and snug it into place. Save the old jet in case you decide to go back to stock jetting.

a flathead screwdriver and count the number of turns as you screw the idle jet's needle valve into the carburetor until it touches bottom. Do not tighten the valve against the jet or you may damage both the needle and the jet itself. Write down the number of turns you counted in case you someday decide to return to stock jetting. Now, unscrew the needle valve the number of turns specified by your jet kit.

Turn the carburetor upside down. Remove the float bowl with either a Phillips screwdriver or Allen key. (A drill with a screw attachment really speeds things up—particularly for bikes with four mixers.) Clean out any gunk that may have collected in the bottom of the float bowl. Also check to be sure that the float bowl gasket is not cracked, dried, or ripped. Replace it if you suspect it is deteriorating.

A float bowl gets its name from the plastic float that controls the fuel height within the carburetor. When working on the main jet, make sure you don't put any pressure on the float, or you could bend the tab responsible for the fuel height. While most jet kits don't have you fiddle with the float height, some do. See the side bar "Setting Float Height," for the lowdown on how to check and set float height.

Turn your attention to the brass main jet located in the center of the carb body. You can identify it by the flathead screwdriver slot in the base of the jet. Make sure your screwdriver fits snugly into the jet or you may mangle it as you attempt to remove it. If the emulsion tube (the brass part into which the main jet is screwed) turns with the jet, hold it in position with an 8-mm wrench. Jet kits without a new emulsion tube simply require that you screw the new jet in place of the old one. If your jet kit includes a new pilot jet, you'll want to replace it the same way you did the main jet. The pilot jet is significantly smaller than the main jet and located near the main jet. Simply unscrew it and replace it with the new one. Before you reinstall the float bowl, make sure that the gasket surfaces are clear of any dirt or grit. One grain of sand can cause a leak that forces you to tunnel back down to the carbs. When screwing the float bowl back onto the carb, make sure the screws are tight, but not overly tight.

Flip the carb so that you now have access to the top cover and unscrew the retaining screws/bolts. When you remove the cover, you'll be presented with the top of a diaphragm. Constant velocity (CV) carburetors use the vacuum created as air flows through the carb's throat to lift a slide. The slide itself performs two duties. First, by restricting the size of the throat opening at lower air speeds (usually during lower rpm), the slide keeps the air speed high as it passes over the nozzle (or emulsion tube), thus atomizing the

Some jet kits include an emulsion tube. Remove the emulsion tube with an 8-millimeter socket as you did with the main jet. Don't over tighten the emulsion tube when you install it.

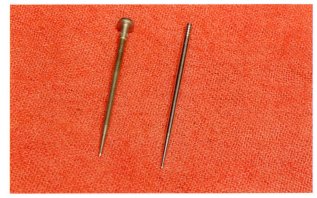

The OE needle (left) has no adjustability, unless you want to shim it up slightly with a washer. The Factory Pro Tuning needle (right) allows you to richen or lean out the mixture as your engine setup requires.

Carefully check the diaphragm on top of the vacuum slide. If the diaphragm has any cracks or pinholes, it will need to be replaced.

fuel more effectively. Second, it helps to keep the engine from stumbling when you whack the carb open, since the slide will only rise as quickly as the airflow requires.

Carefully, lift out the slide and examine the rubber diaphragm. If you find any cracks or pinholes, the diaphragm must be replaced. Notice how the needle hangs out of the bottom of the slide. Press up from the tip of the needle to remove it from the slide—but be sure you remember the order in which parts come out of the slide. You will find a spacer that holds the slide spring in place above the needle. You may also find a small washer below the needle. Carefully set these parts aside for use later.

Since carburetors with factory jetting usually have non-adjustable needles, most jet kits include an adjustable needle with instructions on how to correctly set the needle height. When affixing the circlip to the appropriate slot in the needle, always count from the top slot. If your jet kit instructions specify installing a washer, make sure it is on the needle below the circlip to raise the needle halfway between two circlip notches. If the jet kit requires the slide vacuum hole to be drilled, make sure you deburr the new hole. Slip the new needle and washer (if required) into the slide assembly. Place the spring holder in with its prongs down toward the needle.

When reinstalling the slide, make sure that the needle fits into the top of the emulsion tube. Carefully place the lip of the diaphragm in the groove at the top of the carb body. Any folds or buckles will lessen the vacuum inside the top of the carburetor and prevent proper slide function.

Once you have replaced the jet, installed the needle, and adjusted the idle screw, you need to reinstall the carburetor. A little WD-40 on the carb mouth will ease its reinsertion into the boot. Tighten the clamp to secure the carb. Reinstall the throttle cables and adjust the throttle freeplay. When the engine has warmed up, adjust the idle speed to factory specifications. If the jet kit you installed was designed for your bike's exact engine/pipe configuration, you should now go out and enjoy your new power. However, if you've made other modifications to the engine, you may have some debugging to do in order to make it perfect. Book some dyno time to sort things out quickly.

Setting Float Height

If your jet kit recommends a specific float bowl height, you'll get the best performance from your motorcycle if you take the time to carefully set the height. Your reward will be crisper carburetion in the bottom end with an exhaust note that doesn't sound soggy (too rich) or raspy (too lean).

You'll perform the height adjustment after you've installed the other parts of the jet kit. With the carb standing on end (resting the end on your work bench will make it easier to steady the assembly while you're measuring the float height), tilt it until the float shifts toward the carburetor body. The idea is to have the float move the valve pin to a closed position—and no farther. You don't want the valve spring to compress.

While holding the carb in position, measure the float height with Factory Pro Tuning's float-height tool. The two arms of the tool should rest on the float-bowl gasket surface while the measuring bar gets slid into position so that it just barely touches the highest point on the float (when measured from the gasket surface). You may have to perform several adjustments of the measuring bar to keep it from compressing the float spring slightly—especially if your hands shake like mine.

To change the height, you need to bend the tang over the valve's spring loops. A little bend goes a long way, so be careful. You'll find that it's better to move in little steps rather than trying to make large corrections. Your goal is to get the measurements as close as possible to the desired height with a maximum of 0.5 mm difference among all the carburetors if you have more than one.

Although it may not look like much, Factory Pro's float-height tool is precision-machined to allow accurate measurement of float height. It's worth every penny when you're tuning carburetors.

PROJECT 24

Installing an Aftermarket Carburetor on a V-Twin

Time: 2 hours to days, for custom applications

Tools: Wrenches, sockets, ratchet, screwdrivers, pliers, rags

Talent: ★★ to ★★★★

Tab: $$$$

Parts: Patrick Racing (42 mm or 45 mm) Road Star kit or Keihin/Mikuni carburetor from Sudco, air filter (for custom applications)

Tip: Carefully route the throttle cables so that they do not bind when you turn the bars.

Benefit: More power, cool intake honk!

Although most manufacturers have made the switch to EFI on their big twins, enough carbureted cruisers still roam the boulevards to make installing a higher-performance mixer a popular modification. Since there are tons of big Vulcans and Road Stars huffing through their stock carburetors, you can order a bolt-on kit from Patrick Racing for your Road Star or adapt the same model Mikuni flat slide carburetor from Sudco to fit your model cruiser. The instructions included here deal directly with the Patrick Racing 42 mm Mikuni HSR kit with notes concerning custom applications.

Why would you want to add a flat slide carburetor to your V-twin cruiser—even if it's relatively close to stock?

According to Nigel Patrick, the brains behind Yamaha's factory Hot Rod Cruiser drag race team, replacing the CV carb offers the most bang for the buck of any modification. Patrick says his carburetor kit will bump a stock Road Star's output by 12 horsepower! How? You achieve better airflow into the engine without changing the diameter of the carb throat with a flat-slide carb. CV carbs have a butterfly valve in the center of the throat that still partially obstructs the airflow at wide open throttle with the slide fully retracted into the carb body. The twist grip on flat slide carburetors directly act on the slide (rather than having airflow raise the slide). So, when the throttle is wide open, the slide disappears completely into the carb body leaving nothing in the throat to disturb the airflow.

Once you have the stock carburetor out of the way, you get a clear view of the intake tract. If you buy the same diameter flat-slide carburetor body from the same manufacturer that produced the OE CV carb, the spigot should fit in the original boot.

Look at the size of the carburetor compared to the boot and you can understand how much the boot can flex from the forces exerted on the bike by the road.

(Remember, more air into the engine means that a bigger charge can get into the cylinder.) Another benefit is ease of tuning as you upgrade components on your engine. CV carbs, because they depend on airflow to raise the slide, are influenced by engine back pressure, making tuning more of a challenge. Also, most aftermarket flat slide carburetors allow you to change jets with the carb in place on the bike.

Since you're going to mount a new carburetor, you shouldn't be surprised that you begin by removing the tank, air box, and any crud associated with the stock carburetor. Before you remove the OE carb, take note of what hoses perform what function. Also, note the routing of the throttle cables (if you forget, the factory service manual will have a diagram). Removing the throttle cables from the carburetor can be tough in the cramped space of the engine's V, so you can make the job easier by disconnecting the throttle cables from the handlebar and twist grip. You may also need to remove the choke cable and free up the idle speed adjuster from its mount on the engine. Loosen the hose clamp securing the carburetor to the boot and wiggle the carb free. Drain the float bowl before you set the carb down so you don't make a mess.

Before mounting the carburetor, some mechanics prefer to attach the cables. Give the carburetor spigot and the manifold boot (AKA rubber flange) a spritz of WD-40 to ease their joining. Make sure the clamp on the boot is in a position that will allow you to tighten it when the carb is in position. Insert the spigot into the boot until it snugs in place completely. Check its fit by rocking the carb back and forth lightly. The carb should stay put. Check that the float bowl is level front-to-back before tightening the clamp. Finally, secure the carb by pressing the billet support bracket onto the front of the carb and bolting it to the engine. Lubing the O-ring with WD-40 will make this step much easier.

Those grafting a carburetor to their V-twin without the luxury of an off-the-rack kit should seriously consider taking the time to fashion a carburetor bracket. As you were removing the OE air box, you may have noticed that its backing plate was pretty beefy and attached to the engine in two or three places. This is because the plate—not the boot—is actually supporting the carburetor. Bike manufacturers (and quality aftermarket companies) secure the carburetor for a variety of reasons. First, some bikes have components that the carburetor could hit if it's flapping around on the end of the rubber boot. For example, the Road Star has the fuel gauge sensor protruding from the bottom of the tank above its carburetor. Not only could this rapping damage the expensive sensor, but also a hard enough impact could potentially cause a fuel leak (right above the hot engine). We can all agree that catching on fire while riding your bike would be a Very Bad Thing. This excessive shaking of the carburetor can also cause foaming in the float bowl and hurt performance.

Before mounting the air filter backing plate and air filter, you should hook up and secure all fuel and vent lines. Don't forget to run the throttle cables and adjust them for proper freeplay. Also, make sure that the choke and idle speed controls are hooked up and functional. (As you're mounting the air cleaner's backing plate, pause for a moment and marvel at how the opening for the carburetor throat has been machined with a radius—just like a velocity stack—to smooth the airflow into the carburetor.) If the filter did not come pre-oiled, prepare the filter material according to the instructions from its manufacturer. While holding the filter in position, slide the filter cover in place and tighten its mounting nuts. Check the filter for proper seating before buttoning up the rest of the bike. Go for a ride and enjoy the newfound power.

Of course for *optimum* power delivery, you'll want to book some dyno time . . . .

Inside this bracket, the O-ring offers a flexible-but-firm support for the Mikuni carb.

It's unfortunate that people will only see the shiny billet air filter cover. The craftsmanship of the entire package is worthy of display. Barnett makes all the cables for the Patrick Racing Kit. *Photo courtesy of Patrick Racing*

PROJECT 25 | EFI Tuning

Time: 1 hour (installation), many hours developing maps

Tools: Basic mechanics tools, (laptop) computer and assorted cables for Power Commander, dyno

Talent: ★★★

Tab: $$$$

Parts: Power Commander, Cobra Fi2000, or Dynatek FI Controller

Tip: For a good starting point, use fuel maps created for your exhaust system, if available.

Benefit: The best possible fuel mixture and power at any rpm

The switch from carburetors to fuel injection on cruisers has opened a whole new door to controlling mixture throughout the rpm range. Gone are the days of jetting's black magic. Instead, we've entered the era of electronic hocus-pocus. With ability to adjust jetting infinitely over the rev range of the engine, fuel injection and its associated maps have also given us an infinite number of ways to screw things up. Fortunately, companies such as Dynojet and Cobra have created magic black boxes that can share the engine management duties with the OE boxes of a similar color and purpose.

Installing a Power Commander is about the easiest bolt-on (or is it plug-in?) process possible. Make sure the ignition is off. Then remove the seat to gain access to both the battery and the wiring harness. You may also need to remove the battery cover. Unbolt and remove the tank. Locate the main wiring harness on the frame. Follow the wires from the injector rail to the connector on the main harness. Unhook the connector for use with the Power Commander.

Different cruisers will have different mounting locations for the Power Commander itself. For example, on the Vulcan 2000 featured in the photos, the Power Commander will rest on top of the battery between the shock and the tank support. Connect the unit's black ground lead to the negative terminal of the battery. Locate the main wiring harness on the frame. Route the Power Commander's wire bundle along the bike's wiring harness towards the front of the bike. While you're running the cables, make sure you keep it clear of any places where it might get pinched by the gas tank, seat, or any of the parts you removed. Follow the wires from the injector body to the connector on the main harness. Unhook the connector and attach these to the associated plugs on the Power Commander. Isn't that easy?

If your Power Commander model will be physically mounted to some part of the bike, be sure to clean the location

Snugly tucked away in the back of the trunk, the Power Commander leaves plenty of room for access to its USB port without compromising your storage.

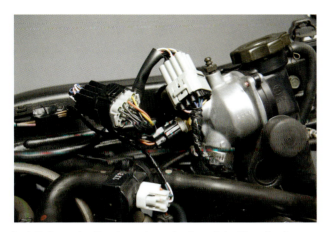

Installations don't get much easier than this. Plug the two black connectors into the wiring harness and select the proper white connectors. Use zip ties to keep everything in place.

Zip tie the Power Commander's wires to your bikes harness.

with alcohol to remove any grease that might prevent the mounting tape from sticking. Place the Power Commander in a position that will give you access to the accessory and USB ports. If your cruiser model requires that you slip a protective cover over the Power Commander, do that now.

Button up your bike so that you can ride it. Running your bike with the Power Commander is now as easy as starting the engine, since you bought it with the base map already installed. If you want to modify the base map, you have two choices. First (and most flexible), hook your Power Commander up to your computer and either manually tweak the maps or download a map from the Dynojet website or your exhaust pipe manufacturer's website. If you're on the road and want to make some adjustments (or don't have access to a computer), you can make mixture changes with the three buttons on top of the Power Commander unit.

With the engine cut-off switch in the run position, depress and hold all three buttons on the Power Commander. Turn on the ignition and wait for the green light to flash on and off. (Note: On some bikes the fuel injection system powers down if the engine isn't running, requiring a special adapter and a small battery to power the unit.) Choose which rpm range you want to adjust and press the appropriate button once. The LEDs on the unit should show the mixture level. If you press and hold the same button, the mixture will lean out, causing the fuel lights to move down. Pushing and releasing the button repeatedly will richen the mixture and move the lights up the scale. When the two LEDs on either side of the 0 light up, the unit is using its unmodified map.

You don't need to worry if you're not comfortable making changes to your Power Commander's maps. Many bike shops that have dynos also have computers and all the appropriate cables to transfer (or develop) maps to your bike. The beauty of a system like this is that as you modify your bike further, you can create new maps to fit your engine's needs. Also, if you don't like the changes you made to one map, with the push of a button, you can revert to your previous one. How about that for twenty-first–century technology?

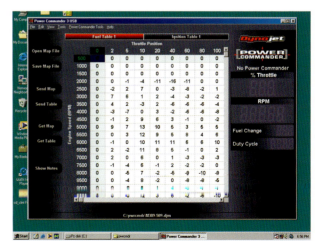

Detail-oriented folks will be in heaven with the Power Commander software. You can twiddle with the fuel and ignition maps to your heart's content. If you're of a Mac persuasion, the application runs fine under Virtual PC.

Cobra FI 2000 and Dynatek FI Controller

Cobra and Dynatek took a slightly different—but still effective—approach to the tweaking of your Cruiser's EFI. Both modules utilize three small knobs (called pots—short for potentiometers) to adjust the fuel mixture. Consequently, no computer or maps are required to adjust your cruiser's tuning. Instead, the three pots control different aspects of the mixture. The primary difference between these modules is that the Dynatek unit both richens and leans the fuel mixture, clipping on with factory connectors, while the Cobra unit uses wire taps and can only richen the mixture.

Installation varies with the model cruiser being used (there were 20 models supported by Cobra and 15 by Dynatek at the time of this book's printing). On the Cobra unit, you essentially connect four wires (two control wires, one power, and one ground). The wires you will be connecting to have the same color coding as those on the Fi2000. Using wire taps that splice into the stock harness without actually severing any wires, you create a link to the Fi2000 that then modifies the fuel mixture based on rpm. The Dynatek simply hooks in to the stock connectors (like the Power Commander described in this project), plus positive and negative leads.

Adjusting either can be done on a dyno or the street—without a tether to a computer. Start from the manufacturer's recommended bike-specific settings found in the instruction manuals and focus on the various rpm ranges one at a time. Of course, spending a little time on the dyno will yield the best results.

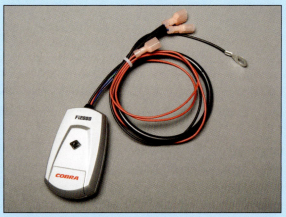

Just a little bit larger than your typical car alarm remote, the Fi2000 allows you to richen three areas of your bike's fuel mixture.

The Dynatek FI Controller utilizes factory wiring harness connectors and allows you to both richen and lean the mixture of your cruiser.

Here they are in all their glory! A jewelers screwdriver is all you need to tune your cruiser.

PROJECT 26 | Dyno Tuning CV Carburetors

Time: 2 hours to several days

Tools: Wrenches, sockets, ratchet, Phillips screwdriver, flathead screwdriver, Allen keys, float-height tool, pliers, drill, rags

Talent: ★★★★

Tab: $ to $$$

Parts: Jets of various sizes

Tip: Finding the best main jet is only the beginning—be patient and finish the job.

Benefit: A bike that carburets perfectly throughout the rev range

Cruiser owners frequently mention having their modified bike tuned on a dyno, but many people don't really know what the process entails. Fuel injection has made the process much easier (and quicker) for those who have a Power Commander since the same computer that is acquiring the information from the dyno also runs the software to alter the fuel injection curves. Owners of bikes with Cobra's FI2000 or Dynatek's FI Controller will still have to do a little fiddling by hand, but the process is still much easier than the process of fine-tuning a carburetor's jetting. Tuning fuel injected requires fewer runs and much less wrenching than working with jetting. While this project is directed towards carburetors, owners of fuel injected bikes will learn much about the process of dyno tuning.

If you've made any changes to your carbureted bike that aren't accounted for in an off-the-rack jet kit (such as bumped up compression or hotter cams), you'll find that your bike seems to work well . . . or it may run like crap. Regardless, you won't be getting optimum performance from your constant velocity (CV) carburetors until you put your bike on the dyno and do some fine-tuning. Marc Salvisberg, owner of Factory Pro Tuning (www.factorypro.com) in San Rafael, California, has developed jet kits for just about every bike made and devised a technique for dyno tuning CV carburetors—a technique that is widely acknowledged to be the most thorough and user-friendly. Although you can perform some of the process with the sensors in the seat of your pants, the best results can only be found utilizing an eddy-current dyno and an exhaust gas analyzer.

Dynamometers can be broken down into two general categories, inertial dynos and eddy-current dynos. The dynos found in many motorcycle shops are inertial dynos. The bike's rear tire rests on a large drum, and the bike's horsepower output is measured by how quickly the mass of the drum can be accelerated as the bike's engine runs in a high gear, with the carbs wide open, from low rpm to redline. While inertial dynos are good tools for the type of work most motorcycle dealerships perform (measuring the output of an engine, generating information tuners need to maximize a bike's performance, or verifying that changes worked), inertial dynos have one drawback. Inertial dynos run sweep (or acceleration) tests in which the readings are taken as the engine makes a run through its rpm range. An entire run through the rpm range on an inertial dyno can take as little as 10 seconds, once the bike is running in top gear.

Eddy-current dynos still place the bike's rear wheel on a large rotating drum, but the dyno gathers information in steps. The engine moves through its rpm range, pausing at regular intervals for several seconds to have the power output measured. At each step, the engine is still running wide open, like with an inertial dyno, but the rpms are held constant by the dyno's electromagnetic brake. The power output, as determined by the amount of force required to hold the engine speed constant, is then measured and recorded up to 30 times per second, providing a very accurate picture of what is happening. Aside from gathering extremely accurate data at each step, holding the engine at each rpm step allows the carbon monoxide (CO) output to be measured, giving the tuner one more piece of data hinting at why the engine does or doesn't perform well at a particular rpm.

Salvisberg, who manufactures and sells both inertial and eddy-current dynoeddy-current dynos, cautions us not compare eddy-current and inertial dynos in terms of which is superior to the other. Instead, consider them different tools designed to fulfill different needs, and the eddy-current dyno is designed to gather and sort through the extensive data

No, you won't have one of these sitting in your garage, but the time and effort spent with your bike on a dyno will pay off.

required for high-end R&D work. Also, since the numbers generated on dynos built by different companies will give slightly different results (due to different manufacturing and calibration specifications), simply comparing dyno numbers without knowing their origin will prove less revealing than comparing baseline and subsequent runs on the same dyno, no matter which type.

From a Baseline Onward
Before you start fiddling with the carburetors, you need to perform baseline dyno runs to determine where your power output is before you make changes. With the baseline data collected, the development process begins, with repeated dyno runs after the carb undergoes incremental, duplicable changes. Every change in jetting creates a change in output, giving some new knowledge about the engine (even if the change made the engine run horribly). You should meticulously record all information, including if the engine misses or backfires. The resulting data is then compared with the baseline and previous runs to determine the next adjustment. Salvisberg stresses that what looks like fooling around—changing the main jet, running the bike on the dyno, and changing the main jet again—is really the way he learns the character of an engine. By concentrating on only the main jet, you make changes with broad strokes (literally moving the fuel delivery curve—the rate at which fuel is delivered to the engine—up and down to see what happens) that gradually narrow down on which main jet produces the best peak horsepower. A nice side benefit of these runs (if you're using an exhaust gas analyzer, that is) is that you can also find what exhaust gas readings are associated with the best power delivery throughout the rpm range, making it easier to tune the low- and midrange later.

Once you've found the main jet that makes the best peak power, only then can you begin to address the needle. Why? In order to visualize how the main jet and needle can vary the mixture at different engine speeds, a few basic carburetor concepts must be understood. Air flowing through the carburetor's venturi flows over the discharge nozzle, whose bottom opening rests below the surface of the fuel contained in the float bowl. The low pressure created by the airflow across the discharge nozzle draws the fuel up through the main jet's orifice to be vaporized by the high-speed air. The needle is attached to the slide, which moves up and down depending on the volume of airflowing into the engine, to keep the airspeed high across the discharge nozzle. As the slide moves up and down, the needle regulates the amount of fuel allowed to vaporize into the airflow. The main jet limits the maximum amount of fuel to enter the airflow at wide open throttle at high rpm, the only place where the slide will be raised to its highest position out of the airflow (i.e., where the engine's maximum power output will be generated). At all other engine speeds, the needle plays an important part in metering the fuel flow. At low engine speeds, the slide is not raised very far, making the diameter of the needle of primary importance. A wider diameter will allow less fuel out of the discharge nozzle than a narrower needle. Where the needle's taper begins, and the shape of the taper, can richen or lean the mixture at midrange rpms.

Working the Middle

So, you've installed the best main jets for your engine setup. Now that your top-end power has been sorted, it's time to focus on the midrange power. What you want is the needle shape and height that will give you the most power in the middle third of the rpm range at full throttle. Since the shaping of needles is a very esoteric skill, you'll be better off ordering a needle from an aftermarket company such as Factory Pro. Your job will be much easier in that you will only be adjusting the needle's height. Starting with the clip in the slot recommended by its manufacturer, make a series of dyno runs in which you either raise or lower the clip. If you raise the clip in the needle for one run and the power in the midrange improves, raise the clip and make another run. Keep doing this until you get the best power at full throttle in the middle third of the rpm range. If, after your first run, the power drops, try lowering the clip below the starting position and continue testing for the best power.

While you could quit work on the midrange at this point, those who want to make sure that their bike makes the best power it can will also check to see if moving the needle up or down one half step will make a positive change. After making a run with a washer under the clip of each needle, you can stop if the power improved. If it didn't, try lowering the clip one slot and placing a washer under it. Make one last run. If the results improve, leave the needles as they are. If they don't, remove the washers and move the clips back up one slot.

Winding Up at the Bottom

At this point, you have the best top-end power and strongest midrange you can get out of your current engine setup. Now, you need to finish up with getting the most power you can out of the bottom end. You're entering an area where few novices ever tread. Consequently, you'll see riders struggle to launch their bikes in traffic. According to Salvisberg, you need to get the engine to accept full throttle in second gear from 2,000 to 3,000 rpm without retching, coughing, spitting, or stumbling. To tune this range to its optimum, you will be listening to the engine as much as you'll be looking at the dyno chart. You'll also need to use a little deductive reasoning to figure out what's going on in the cylinders.

If the engine note is thick, sounding like it has phlegm in its throat that it's trying to clear, and the problem only gets worse as the engine warms up, it is running rich and the fuel level in the float bowls must be lowered. Try lowering the fuel level by increasing the float height 1 millimeter. (If this sounds like a foreign language to you, take a side trip to Project 23 and read the sidebar "Setting Float Height.") However, if the engine sounds dry and weak with a flat spot somewhere in the bottom third of the rpm range, it is most likely lean. Try raising the fuel level 1 millimeter. Of course, if you've never set the float height on your bike, you should check to make sure that it is within factory specs, anyway.

You're almost done, but not quite yet. You should be familiar with the pilot jet circuit if you've ever installed a jet kit. (Remember the plug you had to remove to access the idle screw?) Begin by setting the screw to the number of turns out from full closed recommended by your factory manual or jet kit instructions. A properly adjusted pilot circuit will prevent rough idling and lean surges at steady state cruising at low rpm. Surprisingly, lean surges at high rpm with small throttle openings can also be the result of a pilot problem. For both of these symptoms, richen the mixture by unscrewing the adjuster screw in one-half-turn increments. The tricky part of adjusting the pilot circuit is that the screws on some carburetors control the flow of fuel in the pilot circuit (fuel screws) while others (air screws) control the flow of air. How can you tell the difference? The more common arrangement locates the fuel screw between the main jet and the engine while an air screw is between the main jet and the air box. So, richen your mixture by either allowing more fuel or less air, depending on your carburetor type.

Two other symptoms to look for: When blipping the throttle, if the rpm drops down below the idle speed and slowly climbs back up, the mixture is too rich. (These symptoms may also get worse as the engine gets hot.) Screw in the adjuster in half-turn increments until the problems disappear. If the rpms stay high before settling back down to idle speed after the throttle is blipped, it's a sign of a lean pilot mixture. Unscrew the adjuster in one-half-turn increments. If you can't make the problems go away, you may have a leak in the intake tract you need to hunt down.

Properly tuning your carburetors can be a tedious, time-consuming process. The dividends it pays will be felt every time you throw your leg over your bike. When your friends are bemoaning the flat spots in their jetting, you can simply smile knowingly and tell them how to fix the problem.

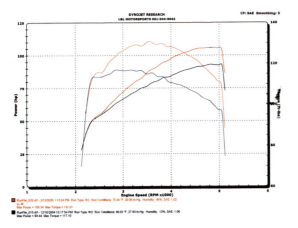

Keep meticulous records while you've got your bike on the dyno. Otherwise, you won't know what changes produced the best power. You need to work in a methodical, organized manner to get the best results in the least amount of time.

PROJECT 27

Installing an Aftermarket Air Cleaner

Time: 1 hour

Tools: Wrenches, sockets, ratchet, Phillips screwdriver, flathead screwdriver, Allen keys, needle-nosed pliers, circlip pliers, rags

Talent: ★

Tab: $$ to $$$

Parts: Baron Custom Accessories Big Air Kit

Tip: You will need to rejet or install a new EFI map with this modification or you may actually reduce your engine's performance.

Benefit: Freer breathing intake, sexy carburetor honk

Complementary Modifications: Install aftermarket pipes (see Project 21)

Air filter kits that remove the stock air box are one of the most popular performance modifications (probably second to the aftermarket pipe) riders make to metric cruisers. The reasons are simple: The chromed billet air filter covers and the pleated fabric filters themselves look really cool, and that's before you consider the increase in performance and the sexy intake honk. Another reason that these modifications are popular is that the actual installation is quite easy, giving novice mechanics a chance to gain some low-stress experience. This project features a Baron Custom Accessories Big Air Kit, but you can find filter kits in a variety of styles from numerous vendors.

One note, though: Because the filter will breathe so much better than stock (particularly with an aftermarket pipe installed), the fuel mixture will be leaner than before. While all filter installations may not require rejetting or adjusting the EFI, you'll want to have the mixture set to get maximum performance out of your bike—even if it doesn't sound like it's running way too lean.

You'll begin by removing the tank (and all the parts you need to pull off to access the tank). Then carefully remove the stock air filter cover. It is usually held in place by screws or Allen bolts. Next, remove the filter itself, which may be screwed or clipped into place. Unmounting the air box backing plate can be a little tricky because sometimes fasteners are tucked away, almost out of sight. If you follow the rule of not using force when a part doesn't simply pull free, you'll be fine. Most bikes will have a place where the hose(s) from the crankcase

You may not be able to remove all the hoses from the OE backing plate until it is free of its mount. The springs and hose fittings tend to be pretty tight.

The air temperature sensor is particularly delicate. Be extra careful when installing it on the new backing plate. Here a circlip is used in conjunction with rubber washers to hold it in position on the new, thinner backing plate.

Tighten the bolts on the backing plate in a crisscross pattern. Make sure that any bolts in the groove for the air filter are completely seated, leaving room for the filter.

breather vents into the air box as well as a place where the exhaust-air-induction system draws air from the air box. Cap the induction system at the cylinder heads with the included rubber caps and secure them with the stock retainer. (Note: Space is usually tight by the heads, so you may have to loosen some parts such as the radiator to cap the hose mounts.) The hoses for the breather will be attached to the new backing plate. On the Vulcan 2000 shown in the photos, the solenoid and the hoses for the remainder of the induction system were also removed (and placed in a strong round container for safe keeping), with the wiring zip-tied safely to the main harness.

Fuel-injected bikes will have an air temperature sensor inside the air box that will need to be moved to the new backing plate. These are often made of plastic, so work with it carefully to avoid damaging it. Slip on the two rubber washers and secure the sensor with the circlip. Bolt the backing plate to the stock part's mounting points or the new mounting bracket (if included in your kit). Secure the crankcase breather to the backing plate. Don't forget to install the hose clamp or spring to keep it from popping off.

Hold the air filter in position, making sure that its edge fits into the groove in the backing plate. While you hold the filter in place, thread the bolts attached to the cover through the corresponding holes in the backing plate. Secure the cover to the backing plate with the washers and locknuts. Before you completely button up your bike, make a final check to see that the filter is securely in position and no gaps are apparent. Once you have this installed, you will have dramatically altered the intake system's ability to draw in air, leading to lean fuel mixtures. You'll need to invest in some dyno time to rejet your carb or develop a new EFI map for your Power Commander or other injection modifier.

You can buy covers for the filter in a variety of styles, from the basic smooth chromed billet (shown here) to flames to custom engravings. It's all up to you.

PROJECT 28 | Installing a Dyna 3000 Ignition

Time: Minutes

Tools: Basic mechanics' tools

Talent: ★

Tab: $$$$

Parts: Dynatek Dyna 3000 Ignition

Tip: A quick series of dyno runs will tell you which setting works best for your engine.

Benefit: More power regardless of the state of tune of your engine

Owners of carbureted cruisers may feel a bit left out with all the attention going to the black boxes for the current, fuel-injected bikes. Well, your salvation is at hand! Dynatek has developed ignition systems for many popular, carbureted cruisers. These black plastic (or blue aluminum, in some cases) boxes replace the factory black box. The Dyna 3000 system includes eight advance curves for various states of basic engine modification. You also have control over the rev limit—a boon to Road Star owners who can't abide the stock 4,200 rpm rev limit. People who have bumped up their engines with turbochargers or big nitrous systems will appreciate the ignition retard modes. The Dyna 3000 also utilizes all factory sensors and mounts in the location of the stock black box.

In reality, installing the Dyna 3000 takes seconds. Getting to the location of the OE black box may take a minute or two, though. Once you have uncovered the stock box, unplug it from the wiring harness and unscrew it from the frame. Using the dip switches on the back of the black plastic unit (or the dials on the end of the aluminum ones) set your desired rev limit and advance curve choice. Bolt the Dyna 3000 into position and plug it into the harness. Wasn't that easy?

The fun part of tuning with the Dyna 3000 is trying different advance curves. Dynatek recommends that you initially use the curve that is closest to stock. How would you know which one it is? By reading the instructions that came with the kit, of course. Then you either do a series of dyno runs (the most accurate) or, in a location that will allow a long acceleration run up to redline in, say, third gear (the most fun) to compare the various curves. Although you have eight curves at your disposal, in reality the curves are two sets of four. The first set is for engines that are mechanically relatively stock. The second set is for engines that have had their compression bumped and, consequently, begin their advance curves later in the rpm range. So, begin your testing in the first range, the A curves. On each successive run change to the more aggressive curve. If the performance is better, change to the next curve. Repeat this until the acceleration run doesn't improve, then move back one step. If the A curves cause detonation (or you have bumped compression), try the B curves. They end up with the same final advance as the A curves but begin advancing later in the rpm range.

The Dyna 3000's ability to change an engine's rev limit is a powerful tool that can do great good and, potentially, great harm. For example, if you choose to increase your Road Star's redline, do not go any higher than 5,000 rpm with the stock valvetrain. Otherwise, you risk an expensive failure of the valves. If you have any doubts on rev-limit settings, check the Dyna's instruction manual or contact Dynatek's tech support before you make the big change.

Now that you have the Dyna 3000 installed on your bike, you have the flexibility to alter your ignition curves as you progressively modify your cruiser's engine. After all, power corrupts, and maximum horsepower corrupts magnificently.

Although this may not look like much, the Dyna 3000 can help you extract extra ponies from you cruiser.

Changing settings via dip switches is easy. Just make sure you have the listing from the directions.

ENGINE

113

PROJECT 29 | Clutch Replacement

Time: 2 hours

Tools: Sockets, Allen sockets, ratchet, torque wrench (foot-pounds and inch-pounds), picks, contact cleaner, gasket sealer, rags, oil catch pan, grease pencil or marker, scraper, sandpaper, solvent, zip ties

Talent: ★

Tab: $

Parts: Barnett clutch kit (fiber plates, steel plates, and clutch springs), clutch cover gasket

Tip: Fine-tune your clutch engagement with a combination of stock and aftermarket springs.

Benefit: Positive clutch engagement.

Most riders rarely think about their bike's clutch, if at all. Sure, they use it every time they ride. They may even do the right thing by adjusting and lubricating the cable occasionally. But do they ever consciously think about the clutch? Never. That is, until it becomes grabby or starts to slip. (Been drag racing, have you?)

Even if you haven't been drag racing, your bike's clutch wears every time the plates slip over each other as the clutch is engaged or disengaged. At the first sign of clutch failure, you should replace it. Don't wait until your clutch fails completely and takes more expensive engine components with it. The signs of clutch wear include: slipping under power, loss of clutch "feel," grabbiness, or some other marked change in clutch function. Although you can often get away with only replacing the clutch's fiber plates, performance-minded folks replace all the plates and springs to ensure that everything is within specs.

When you order replacement fiber plates, you may be faced with the choice of either carbon fiber or Kevlar friction material. Carbon fiber plates can handle more abuse, which makes them ideal for racing situations. That additional strength used to come at the expense of being slightly more abrasive to the steel plates, wearing them out quicker. So Barnett Tool and Engineering offered the less abrasive (and slightly less durable) Kevlar plates for street riders—since street riders don't abuse and replace clutches as frequently. Now, according to the company, they've refined the carbon-fiber compound to give the same abuse tolerance as Kevlar without the addi-

The importance of carefully setting parts aside can't be overstated. Zip tie the clutch plates together to maintain their order. If fasteners are arranged neatly and logically, reassembly of the clutch pack and covers will be much easier.

The springs play the key role of holding the pressure plate against the clutch plate. Measure the free length of each spring to make sure it is within manufacturer's specifications, or better yet, play it safe and replace the springs along with the clutch plates.

Slip the fiber and steel plates free of the clutch basket and hub with your fingertips. If you can't reach them, use a pair of picks.

tional abrasiveness. Consequently, their entire line of fiber plates will be moving exclusively to carbon fiber in the next couple of years. So, consider carbon fiber the future of clutch materials.

Start by leaning the bike away from the clutch side so that the oil won't leak out when you get the clutch cover off. You'll need to lean your bike over farther than simply using the sidestand. Leaning your cruiser against a wall will do the trick. (If you're replacing the clutch after it failed, you should change the oil and filter since they're most likely contaminated with clutch-plate particles.) Next, using the appropriate socket, loosen all the clutch cover bolts in a crisscross pattern. Pick a point on the cover (mark it with a grease pencil if you're forgetful), remove the bolts one at a time in either direction, and place them in order on a clean shop rag. You may find that the bolts vary in length, so maintaining their orientation will be vital on reassembly. Position a pan to catch any oil that may leak out when you remove the cover. Tap the cover along its edge with a rubber mallet or deadblow hammer to loosen the gasket sealer. Pull the cover free. If the cover still won't pull free, locate the pry tabs on the clutch cover and gently pry the cover free of the case with a screwdriver or pry bar.

To clean up the cover, get rid of any remnants of the old gasket with a knife or gasket scraper. Be careful not to score the sealing surface. Sometimes little bits of the sealant will refuse to relinquish their hold on the surface. You can use some very fine (600-grit) sandpaper to vanquish these last bits of crud. Chemical strippers such as naval jelly can help in particularly tough cases, but be careful to keep these chemicals away from all painted parts. Make sure all gasket pieces are removed. Clean the mounting surfaces with a solvent such as contact cleaner to make sure no oily residue remains to interfere with gasket adhesion. Covers with O-rings use no sealant, so pay special attention to this step.

Remove the five or six clutch pressure plate bolts with a ratchet or air wrench. Remove the springs and set them aside with the bolts. (Note: Many Yamaha models have a diaphragm spring instead of several coiled springs. Don't be put off by this. It unbolts the same way that the coiled springs do.) Before removing the pressure plate, take note of any alignment marks on the clutch pressure plate and basket that may need to be matched on reassembly. The throw-out bearing in the center of the basket may fall out as the pressure plate is removed. If it does, check for alignment marks and place it back into position. Using the tips of your fingers or a pair of curved picks, remove the clutch plates one at a time and stack them in exactly the same order. Note the plate order for installation. Incorrectly stacked plates can cause premature clutch failure. (Zip-tying the clutch plates together will keep you from mixing things up.) You may also need to remove an inner steel plate that is secured with a retainer ring on some bikes. Since this plate is usually thicker than the other steel plates, set it aside to avoid mixing the plates.

While the clutch is apart, inspect the clutch basket's inner and outer hubs for wear. If any notches or grooves are visible, the basket may need to be replaced. Installing a new clutch into a worn basket may result in abrupt clutch engagement or clutch chatter.

If you don't plan on replacing the steel plates along with the fiber ones, check the steel plates for any signs of wear, such as discoloration or scoring. Measure the steel plates' thickness to make sure the plates are within the manufacturer's recommended tolerances. Make sure the plates are not warped by placing them on plate glass or another known flat surface. If any of the plates do not lay flush to the surface or can be rocked in any direction, replace the plates as a set.

After you've removed the clutch plates, inspect the inner and outer hubs where the plates make contact. You're looking for any wear that creates indentations in the basket fingers. If you find any, you'll need to replace the basket assembly. Some polishing of the metal by the plates is to be expected.

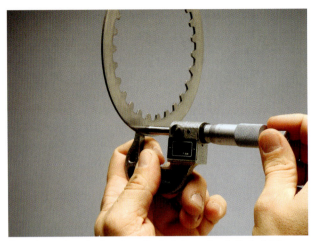

If you're not replacing the steel plates, check their thickness. Inspect them for discoloration or scoring. Finally, make sure they are perfectly flat, using a thick glass surface. Or, save yourself the trouble and buy new ones.

Soak the fiber plates long enough for the oil to completely permeate the porous material. Failure to follow this step will result in a grabby clutch.

Before assembling the new clutch pack, soak the fiber plates in fresh oil for about 5 to 10 minutes. When sliding the plates into the clutch basket, be sure to arrange fiber and steel plates in exactly the same order as the old clutch pack. (This is why zip-tying the old pack together is a good idea.) If you are unsure, the innermost and outermost plates are usually fiber. If your bike's innermost steel plate is of a special thickness or requires a retaining ring, make sure that you install the correct plate. Also, some OE replacement plates may have color coding to specify the order. Consult your factory service manual. The steel plates are usually made of stamped metal, with one rounded edge and one sharp edge. Some mechanics say to make sure the steel plates are installed with the sharp edge facing the pressure plate or excessive outer hub wear may result. Barnett says that all the steel plates should be installed the same way and that it does not matter if the sharp edges face in or out. In fact, the company's vibratory deburring of the steel plates makes the whole "sharp edge" issue a moot point.

Before you reinstall the pressure plate, clean any surface rust or corrosion off the clutch pushrod to guarantee smooth clutch actuation. A thin film of grease between the pushrod and throwout bearing will help things work more smoothly.

Place the pressure plate over the clutch pack. Remember to match up any alignment marks on the clutch plate and basket. Install the springs into the pressure plate and screw the bolts in until snug. Be sure to install the springs and bolts in a crisscross pattern for even pressure on the plate. Using a torque wrench, tighten the bolts—again in a crisscross pattern—to the manufacturer's specified torque. Although the OE springs may show no signs of wear and exceed the minimum length specifications, choosing to replace them with the Barnett springs is cheap insurance that the plate gets held in place. While the Barnett springs are approximately 10 to 15 percent stiffer and may require a slightly firmer pull at the lever, the company says that the additional tension provided by the springs helps make engagement of the carbon fiber or Kevlar plates more progressive. If you find the increased lever effort objectionable, you can always tune the force by using a mixture of the aftermarket and stock springs—just make sure that you install matching pairs on opposite sides of the pressure plate. Barnett also offers a spring conversion kit for bikes with diaphragm springs, giving owners of those bikes the tunability mentioned above.

To remount the clutch cover, apply a thin coat of gasket sealant to both gasket-mounting surfaces. A pliable, non-hardening sealant works best. If you are unsure of where to apply the sealant, look at the shape of the gasket itself. After allowing the sealant to skin over for a couple of minutes, place the new gasket (remember, the $12 you save by reusing the old gasket will seem inconsequential if the cover leaks oil, and you have to take it off again) in position on the engine case. While you're waiting on the sealant, you can install the dowel pins (if any) in the case. The sealant should hold the gasket in position. Reinstall the cover bolts in the same order that they were removed, but do not tighten more than finger tight. Once all the bolts are installed, torque them to the factory-specified setting in a crisscross pattern. Don't forget to refill or check engine oil level.

Let the bike sit for a half-hour or so to allow the gasket sealant to set before taking your bike out for a ride. Your new clutch will most likely engage in a slightly different lever position, and you may need to adjust the clutch cable freeplay. You'll also notice how much more positively it engages when compared to the tired old clutch you removed.

Make sure you line up the punch marks if your bike has them.

Barnett offers a replacement kit that allows the use of coiled springs instead of Yamaha's diaphragm on the Road Star Warrior, and several of the V-Stars. Users of this kit will be rewarded with better lever feel and the ability to tune clutch stiffness with different spring strengths.

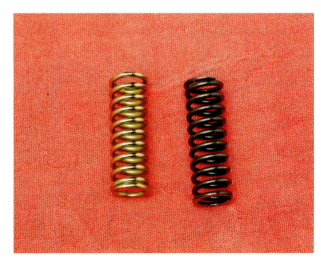

Although you may not be able to see a difference, aftermarket clutch springs (left) are heftier than stock ones. What you'll notice at the lever is a stronger pull. Clutch engagement should be more positive, though.

PROJECT 30

Installing a Big Bore Kit

Time: A day or more

Tools: All your tools, plus ring compressor, piston pin puller, feeler gauges, piston support

Talent: ★★★★★

Tab: $$$$$

Parts: Big bore pistons, rings, replacement (or bored) cylinder blocks, wrist pin retainer clips, cylinder/head gaskets

Tip: Triple-check that you have all the parts you need before disassembling your engine.

Benefit: Performance that can only come from an increase in displacement

Today's cruisers carry displacements that would have shocked riders just a few years ago. Maybe you've got one of those formerly-big twins and feel the need to keep up with your friends. Regardless of what your reasons are, increasing an engine's displacement is the quickest way to gain extra horsepower. If intake and exhaust modifications are designed to get more power out of each explosion in a cylinder, it follows that making those cylinders bigger would allow for even bigger booms per combustion cycle. After all, isn't bigger always better? (We're talking about engines here, not bellies.)

The OEMs have changed the face of big bore kits in recent years. It used to be that you could order up a larger set of pistons for your model bike and take the block to your local machine shop and have it bored out to the proper specifications. However, many cylinder bores are now created directly in the aluminum block and protected with special plating. While this improves cooling and decreases friction in stock engines (both good things), it complicates boring. These blocks require special hones and liner coatings, which means the number of shops capable of matching the tolerances of the factory part—and then plating the bores—is very small. So small, in fact, that you'll find only a handful in the entire country. The other choice is to buy aftermarket cylinders or have your cylinders sleeved.

Riders of those Milwaukee-based bikes have a ton of aftermarket cylinders available. Those of us with metric cruisers will need to get our cylinders bored and sleeved with good old-fashioned iron cylinders. Fortunately, a few aftermarket companies like Patrick Racing (shown in the photos) and Baron Custom Accessories, offer cylinder exchanges.

It's easy to see how much bigger the Patrick Racing piston is next to the stock one. The domed top increases the compression for even more power.

Look at the stock Warrior cylinder on the left. See how the bore is in the aluminum of the jug. The right one has been bored our and had an iron sleeve pressed into it to accommodate Patrick's larger pistons.

Checking the ring's end gap is vital. If the gap is too large, the ring won't seal the cylinder, leading to blow by and loss of power. If it is too small, the ring could destroy the piston and cylinder.

They send you cylinders for their piston kits, and you return your stock jugs to them.

A word of warning about installing piston kits: This project involves many precise measurements and assemblies that, if measured wrong, can leave you with an enormous expensive paperweight. Novice mechanics should only attempt this project by assisting an experienced builder. Because of all the engine- and model-specific steps involved in assembling cylinders and top ends, consider this an overview before you go out and research all that is involved in installing bigger pistons in your specific engine.

Since you'll be digging into the bowels of your engine, you'll need your bike stripped of any parts that will interfere with your access to the engine. So make sure you have a safe place to set aside the tank, bodywork, radiator, EFI, exhaust system and other stuff you'll need to remove. When you're involved in a major project that will most likely take a full day—or more—label all connectors and hoses. If you run in to any complicated cable or wire routings, take a couple digital photos before you remove the parts so you'll be able to reassemble them later. Nuts and bolts should be wrapped in baggies and labeled. Don't just toss them in a bucket, or you'll add many hours to your reassembly.

To replace the pistons, you'll need to disassemble the top end all the way down to the cases. You won't have to split the cases unless your kit requires that they be bored to accommodate the larger cylinder liners. Most kits, however, will not have this requirement. The way you strip apart the top end will depend on whether the engine uses overhead cams or pushrods.

Only for the brave (or experienced): Rings that are too tight can be widened with the careful application of a flat file.

Why risk damaging the piston rings when you can buy a ring spreader dirt cheap. Still, only spread the ring just wide enough to slip over the piston.

Overhead Cams

Begin by putting the front cylinder at top dead center (TDC), as described in your factory service manual. (Of course, removing the spark plugs will make the process of rotating the crankshaft much easier, but you already knew that, didn't you?)

If the engine has an externally accessed cam chain tensioner, loosen it. Internal tensioners should be loosened once they are exposed. You'll then need to remove the cylinder head cover. Pay special attention to any oil lines or filters that may need to be removed.

The cam may be secured by a cam holder or the rocker box by a two-piece construction. Remove the cam holder bolts in a criss-cross manner. Follow your service manual's recommendation for two-piece rocker boxes. Remove the cam. Once the cam is out, wrap a piece of wire around the timing chain to keep from losing it into the bottom end.

Loosen the head nuts in a criss-cross pattern before removing them. Tap the head with a rubber mallet to loosen the adhesion between the head and cylinder then slip the head free. If your engine has pry points, gently use them.

Remove the cylinder in the same manner. Move on to the next cylinder by rotating the crankshaft to TDC for the second cylinder. Don't forget to take up the slack in the timing chain to keep the teeth engaged with the chain and avoid jamming it inside the bottom end.

Now repeat the above steps for the second cylinder.

Notice how the piston is held in position perfectly by a couple pieces of wood. Of course, you can spend money and buy pretty billet ones.

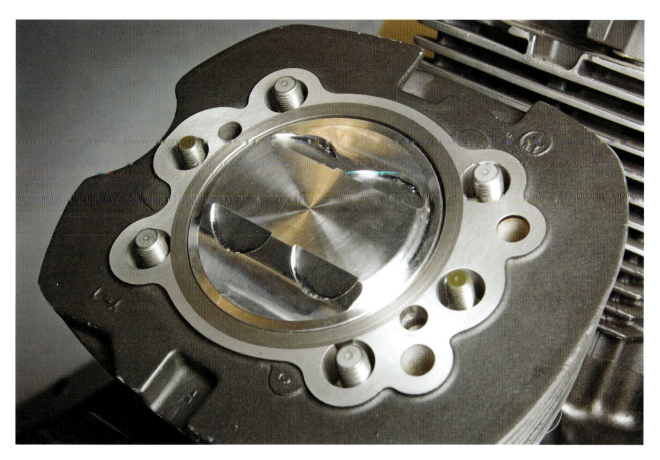

See how the flat of the piston doesn't rise above the top of the cylinder. If it did, it would contact the head when the engine was running—a very bad thing.

Pushrods

Pushrod engines follow the same route, only slightly different. With the front cylinder in TDC, remove the head cover in a criss-cross pattern. Be sure to check your service manual for any sneaky ones hidden out of plain sight—as well as any hoses and dowel pins that might fall into the depths of the engine, requiring you to split the cases.

Remove the rocker arm base and any other components blocking access to the pushrods. Remove the pushrod cover. You should now have access to the head nuts. Loosen and remove these in a criss-cross pattern. If the head is secured by six studs rather than four, you'll want to start that criss-cross on the two middle nuts. Just check your manual if you're not sure. (Note to Warrior owners: The head nuts on the Warrior are not solid caps like on the Road Star. Instead, they have holes in them to save some weight. Unfortunately, this space will fill up with oil, making removal of the head a bit messy. Have some rags ready.)

Tap the head with a rubber mallet and remove it. Once you have the head set aside, free the cylinder with the mallet and slide it free.

Torquing the head nuts in the right order is vitally important. Usually, the order is the reverse of the order you loosened them. Also, torque to the final value in two steps, with the first setting being about half the final one.

Pistons

So now your pistons are standing naked above the bottom end. The very first thing you should do is cover the holes in the cases with clean rags. Once that chore is take care of, you can remove one of the retaining rings (piston pin clips) from each piston. Using a piston pin puller, pull the pin out of the piston and rod assembly.

Warning: Do not attempt to drive the pin out with a hammer! Unless, that is, you want to experience splitting the cases to replace the piston rods.

Congratulations, you're halfway done.

Before you slap your new pistons onto the rods, you need to do some measuring. Even though the rings that came with your cylinders and pistons should have the right gap (when inserted in the cylinder bore), machining tolerances do have some error in them. So, you need to measure the "end gap" in each of the piston rings (except the oil rings) in the cylinder where they will spend their useful life. Following your manual's recommended depth, push the rings individually into the cylinder into which they will be installed. To assure that they are level in the bore, use a piston to slip them into position.

Measure the space between the ends of the ring with a feeler gauge to make sure the gap is correct. If the gap is too tight, you can file it with a flat file to reach the desired measurement. This requires patience and a light touch, remeasuring after every pass over the file. (If a piston ring's ends touch in the cylinder under a load, they will weld themselves together and engine failure will result.) If the gap is too wide, replace the ring. You won't be able to measure the oil ring spacer's end gap, but you will be able to check the oil ring rail.

Using a ring spreader, install the rings in the correct order with their letters (or numbers) facing up. While mechanics have installed rings with their fingers for years, a ring spreader greatly reduces the chance of tweaking the ring as you install it. Consider it to be ten dollars' worth of insurance.

Once you have all of the rings on the piston, you'll want to mount it on the rod. Most pistons will have a mark designating the portion of the piston that should face the exhaust port. Double-check this before inserting the piston pin. Coat the pin with molybdenum disulfide assembly lube before inserting it in the piston and rod. Install *new* pin clips on both sides of the pin. Installing one clip in the piston before pressing in the pin will make things easier. Repeat for the other piston.

Once you have the pistons mounted to the rods, you need to set the ring gaps in their proper orientation. Rule of thumb dictates that they should be diagonal to each other, but motorcycle manufacturers usually have specific orientation requirements for the ring gaps. You should follow those guidelines.

A variety of theories govern the recommended lubricant (if any) to be used on the piston skirts and cylinder walls. Some tuners claim that assembling the cylinders dry will speed the seating of the rings. Others attribute the same result to using only WD-40. Manufacturers tend to be more conservative and will recommend motor oil or molybdenum oil. If you're unsure about what to use, contact the company that produces your piston kit. Otherwise, simply follow your factory service manual's recommendation.

Place the base gasket over the studs and slide it into position on the top of the case. Make sure you also place dowel pins in their holes. To ease the insertion of the pistons into the cylinders, simply support them from below so that they go straight into the bores. You can buy a pair of piston support tools or simply make your own out of a couple of pieces of wood wrapped in duct tape. Slip the supports in place under the piston and turn the crankshaft until the piston rests on them. Don't forget to support the timing chains while doing this.

With V-twins, where you are installing one cylinder jug at a time (rather than several at once, as with an inline four's block), the use of a ring compressor is optional—if you're gentle with the rings as you press them close to the piston and slip the cylinder over them. Slip the cylinder over one ring at a time until it is completely inside the cylinder. If you can't do this with your fingers, try a small flat head screwdriver.

Once you have the piston inside the cylinder, remove the piston supports and seat the jug against the base gasket. Although it's pretty obvious when you're doing it, you'll need to feed the timing chain through its tunnel while installing the cylinder. You may feel like you need a third hand to do it though.

While you may think you're safe to go ahead and reassemble the top end, you've still got one more safety check to conduct before you mount the head. Rotate the crankshaft until the piston is TDC. Check to make sure that the piston flats do not extend above the top of the cylinder. For domed high-compression pistons, this would be the flat ring around the outside edge of the piston top. Now, you can safely install the head gasket and head.

The rest is relatively easy work, with one exception. You need to make sure you set your cam timing to factory spec. With overhead cams, this involves lining the crankshaft up with the appropriate mark, pulling the timing chain tight, and aligning marks on the cam sprockets with the top of the head. You may even have to count links between cam sprockets. With pushrod engines, the marks are on the gears powering the cam. Let your factory service manual be your guide.

Button up and torque the head to the right values and in the right order. Remember that torquing the head is a two-step process. Hook up the cooling, exhaust, and fuel systems. Mount the gas tank and bodywork. Carefully, break in the freshened engine, following the instructions in the accompanying sidebar. And you're done. Easy, right?

You have just completed one of the most difficult jobs to do on a motorcycle.

Breaking it in Right

When you turn the ignition key for the first time on a new or rebuilt engine, you have a window of opportunity to assure that the engine makes the best power it can in the future. Consequently, the method you use to help ensure your bike's future horsepower output can also improve your engine's longevity.

The approaches to breaking in engines are as varied as the owners of the bikes. However, the two extremes of the break-in continuum are represented by the "factory knows best" adherents and the "run it like you're going to race it" supporters. Neither of these approaches is certain to give you maximum horsepower, and one may actually damage your engine. As with many things in life, taking a more middle-of-the-road approach can deliver the power you crave while not prematurely aging the engine.

Piston rings seal the gap between the piston and the cylinder by using the build-up of pressure in the compression and expansion strokes of the piston. This pressure travels through the space between the cylinder wall and the piston to the ring, which holds itself in place due to its natural springiness. When the gas hits the ring, it presses it down into the bottom of the piston's groove, opening a slight gap between the top of the ring and the groove. The pressurized gasses force their way between the top of the groove and the ring and ultimately work their way behind the ring on the innermost point of the groove.

As Kevin Cameron puts it in his book *Sportbike Performance Handbook*, "the gas pressure 'inflates' the ring by...pressing outward against the cylinder wall."

New or freshened cylinder walls have a crosshatch pattern honed into them that serves two purposes.

No synthetic oil during engine break-in if you want your rings to seat properly. Use cheap, high-quality motor oil.

First, the pattern retains oil to help lubricate the rings. Second, when the crosshatches are fresh they have sharp peaks and valleys that need to be worn down—as does the surface of the rings. In order to get the rings to seat against the cylinder wall, you need to run the engine hard enough to utilize the gas pressure forcing the rings against the walls. If you don't use enough force, the rings don't get worn in well enough to form a tight seal against the crosshatch. Use too much force, and heat builds up on the cylinder walls and rings, preventing a good seal. Cylinders with poor ring sealing will never produce the power of which they are capable, since the explosive charge "blows by" the rings, reducing combustion pressure.

Racers and engine builders will often break in engines on a dyno so that they have complete control over the process, but street riders can do a good job of breaking in a new engine if they follow a couple of simple rules. Fill the engine up to its upper oil level mark with cheap, non-synthetic, name-brand motor oil. You want the oil to be of a decent quality, so go with the big names. However, you're going to dump it really soon, so find it on sale. Next, heat-cycle your engine prior to riding it. Start the engine and vary the engine speed to between 1,500 and 2,500 rpm so that the engine spins fast enough to pump lots of oil through. If you're breaking in a rebuilt engine, check for oil leaks while it warms up. As soon as the engine hits operating temperature, shut it down and let it return to ambient temperature. Repeat this process once more. Warm the engine once more, and it's now ready to be ridden.

Avoid riding in a time of day where you will spend excessive time idling or riding at extremely low speeds. You want to be able to accelerate and decelerate without having to worry about other vehicles. Stay out of top gear for now. Out on the road, you want to vary your engine speeds, not holding any one rpm for too long. (One of the worst things you could do is go drone on the interstate at the same rpm for an hour.) Do not lug your engine at any time during the break-in process. Lugging is actually more damaging than excessive throttle at this stage!

When road conditions allow, do some one-half to three-quarter throttle acceleration runs up to about two thirds of redline rpm. When you back off the throttle, let the engine decelerate for at least the same amount of time that you were on the gas. This allows the cylinder walls and rings to cool. If you notice smoke in your exhaust during deceleration, don't worry. Smoking on deceleration is common with a new engine. After riding at varied throttle settings for a couple of miles, repeat the acceleration again. Repeat the process one more time (for a total of three acceleration runs). As you're riding, you may notice that the engine's idle speed has begun to climb with the loosening of its internals. Adjust it to keep it in the 1,500 rpm range for now.

Now your engine is ready for some runs to higher rpm. Keep the throttle at about three-quarters for now. (You'll get to go WFO before too long.) Run the rpms up higher than before (to about three-fourths of redline), and start from about 2,000 rpm. Continue to follow with that extended cooling deceleration. Also continue to vary rpm and speed while avoiding top gear.

After three or four runs up to three-fourths of redline, you're ready for some full-throttle fun. Beginning at 2,500 rpm, crank the throttle all the way to the stop, letting the engine run up to redline. Cool-close the throttle and let engine braking slow the bike down. Since you're running the engine up to the limit, you can generate some serious speed, so be smart about where and when you do this. Also, do it in lower gears, like second or third. Again, you'll want to repeat these acceleration runs three or four times, with extended cooling in between.

If you follow these steps, your engine should now be broken-in. (Remember, this break-in technique is about the progressive loading of the piston rings against the cylinder walls, not a set mileage.) Adjust the idle speed to the factory specification. Dump your oil and replace it with some more inexpensive, name-brand motor oil. You can switch to synthetic after 1,500 miles.

Some things to remember: Keep varying the engine speed for a few hundred miles and continue to avoid lugging the motor. A smart, conscientious break-in will almost guarantee that your bike will readily accept any modification you make to it.

RESOURCES

Barons Custom Accessories:
(888) 278-2819, www.baronscustom.com

Barnett Tool and Engineering:
(805) 642-9435, www.barnettclutches.com

Cobra Engineering:
(714) 692-8180, www.cobrausa.com

Craftsman/Sears: www.craftsman.com

Denso Sales California, Inc.:
(310) 834-6352, www.densocorp-na-dsca.com

Dynatek:
(626) 963-1669, www.dynaonline.com

Dynojet Reasearch, Inc.:
(800) 992-4993, www.powercommander.com

Factory Pro Tuning:
(800) 869-0497, (415) 491-5920,
www.factorypro.com

Finish Line (see Lockhart Phillips)

Goodridge USA:
(310) 533 1924, www.goodridge.net

Home Depot:
www.homedepot.com

Honda Direct Line:
(888) 258-6699, www.hondadirectlineusa.com

Lockhart Phillips:
(800) 221-7291, (949) 498-9090,
www.lockhartphillipsusa.com

K&N Engineering:
(800) 858-3333, www.knfilters.com

Motion Pro:
(650) 594-9600, www.motionpro.com

Motorcycle Mechanics Institute:
(623) 869-9644, www.trade-school.org

Motorex:
www.motorex.ch

Motul:
www.motul.com

Patrick Racing:
(714) 554-7223, www.patrickracingbillet.com

Performance Machine:
(714) 523-3000,
www.performancemachine.com

Progressive Suspension Inc.:
(877) 690-7411, (760) 948-4012,
www.progressivesuspension.com

Race Tech Inc.:
(909) 279-6655, www.racetech.com

Roadgear Inc.:
(800) 854-4327, (719) 547-4572,
www.roadgear.com

Snap-On:
www.snapon.com

Works Performance:
(818) 701-1010, www.worksperformance.com

INDEX

Air cleaner, installing aftermarket, 110, 111
Air compressor, 13
Air filter, 102
    Cleaning/replacing, 28, 29
Allen wrenches, 9
Antifreeze tester, 26
Big bore kit, installing, 118–122
Bike lift, 39, 47, 55, 74, 83, 86
Bike supports, 11
Brake
    Bleeding tool, 43, 47, 50
    Drum maintenance, 39–42
    Line, installing stainless steel, 47–49
    Pads, 34–38
        Break-in, 38
        Change, 34–38
Breaking in the engine, 123, 124
Cable lubrication tool, 18
Caliper
    Aftermarket installation, 55–57
    Rebuild, 50–54
Carburetor
    Balancer, 96
    Installing aftermarket on a V-Twin, 102, 103
    and EFI throttle bodies, synchronizing, 96–98
    Cartridge emulators,
        Adjusting damping, 82
        Installing, 77–82
Cartridge fork tool, 86
Circlip pliers, 68, 71
Clutch
    Adjusting freeplay, 14–17
    Replacement, 114–117
    Springs, 116, 117
Cobra FI 2000, 106

Coolant, checking and changing, 26, 27
Crush washers, 48–50
CV carburetors, dyno tuning, 107–109
Dial caliper, 39, 40
Disc, aftermarket installation, 55–57
Drill press, 79
Drill, 10
Dyna 3000 ignition, installing, 112, 113
Dynamometer, 104, 107
Dynatek FI Controller, 106
EFI throttle bodies and carburetors, synchronizing, 96–98
EFI, 110
    Tuning, 104
Exhaust system
    Installing full, 90–95
    Manifold gaskets, 90
    Pipes, styles, 94, 95
    Slip-on, 93
Factory service manual, 7, 11
Feeler gauges, 118
Float height
    Setting, 101
    Took, 99
Fork oil, changing, 71–73
Fork seals, replacing, 74–76
Fork springs, installing, 68–70
Hammers, 10
Head gaskets, 118
Hydraulic fluid, change, 43–46
Ignition, installing Dyna 3000, 112, 113
Impact driver, 10
Impact wrench, 55, 77, 78, 86
Jet kit, installing, 99–101
Jeweler's screwdrivers, 68, 71, 74, 75

Lights, 12
Loctite, 58, 77
Lowering, 86–89
    Cautions, 88
    Front suspension, 86, 87
        Cartridge forks, 86, 87
        Damping-rod forks, 86
    Rear suspension, 87
Lubrication, 18–20
Master cylinder rebuild, 50–54
Miter box, 68
Oil, changing, 21–25
    Filter wrench, 21, 23
    Screen, 24
Overhead cams, 120
Parts washer, 12
Pipe/tubing cutter, 64, 68
Pistons, 119–122
    Pin puller, 118
    Support, 118
Pliers, 10
Power Commander, 104
Preload, adjusting, 64–67
Pushrods, 121
Ratio Rite, 71
Ring compressor, 118
Ring spreader, 120
Rotary tool, 12
Sag, measuring, 64–66
Scotch-Brite, 50, 52
Screw removal kit, 12
Shaft drive housing, oil check/change, 25
Shock preload adjusting tool, 64
Shock, installing aftermarket, 80–85
Shop setup, 7

Simple Green, 34
Sockets and drivers, 8
Sockets, T handle, 12
Spark plugs
    Checking and replacing, 30–33
    Reading, 31
    Wires, 32
Speed wrenches, 12
Spring compression tool, 71
Spring puller, 90
Tap and die set, 12
Throttle, adjusting freeplay, 14–17
Toolbox, 10, 11
Tools, 7–13
    Cutting, 10
    Electrical, 10
    Measuring, 10
    Specialty, 13
Torque wrenches, 8
Universal joint, 83, 86
    Sockets, 90
Vacuum bleeder, 44, 46
Vise, 12
Water Wetter, 26
Wheels, installing aftermarket, 58–63
    Belt drive, 60, 61
    Front, 59
    Rear, 59
    Shaft drive, 59, 60
Wrenches, 9

MOTORBOOKS WORKSHOP

The Best Tools for the Job.

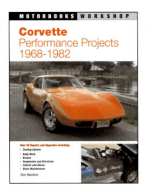 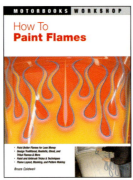  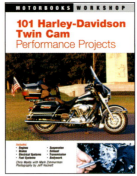 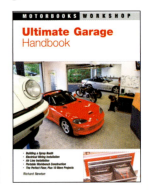

Other Great Books in this Series

Performance Welding Handbook
2nd Edition
0-7603-2172-8 • 139436AP

How To Paint Flames
0-7603-1824-7 • 137414AP

**How To Build
Vintage Hot Rod V-8 Engines**
0-7603-2084-5 • 138703AP

**Honda & Acura
Performance Handbook
2nd Edition**
0-7603-1780-1 • 137410AP

**Hot Rod
Horsepower Handbook**
0-7603-1814-X • 137220AP

**How To Build the Cars of
*The Fast and the Furious***
0-7603-2077-2 • 138696AP

**How To Tune and Modify Engine
Management Systems**
0-7603-1582-5 • 136272AP

**Corvette Performance
Projects 1968–1982**
0-7603-1754-2 • 137230AP

Custom Pickup Handbook
0-7603-2180-9 • 139348AP

**Circle Track Chassis
& Suspension Handbook**
0-7603-1859-X • 138626AP

**How To Build A West Coast
Chopper Kit Bike**
0-7603-1872-7 • 137253

**101 Harley-Davidson Twin-Cam
Performance Projects**
0-7603-1639-2 • 136265AP

**101 Harley-Davidson
Performance Projects**
0-7603-0370-3 • 127165AP

**How To Custom Paint
Your Motorcycle**
0-7603-2033-0 • 138639AP

101 Sportbike Performance Projects
0-7603-1331-8 • 135742AP

Motorcycle Fuel Injection Handbook
0-7603-1635-X • 136172AP

**ATV Projects: Get the Most Out
of Your All-Terrain Vehicle**
0-7603-2058-6 • 138677AP

**Four Wheeler
Chassis & Suspension Handbook**
0-7603-1815-8 • 137235

**Ultimate Boat
Maintenance Projects**
0-7603-1696-1 • 137240AP

**Motocross & Off-Road
Performance Handbook
3rd Edition**
0-7603-1975-8 • 137408AP

**How To Restore Your
Wooden Runabout**
0-7603-1100-5 • 135107AP

Ultimate Garage Handbook
0-7603-1640-6 • 137389AP

**How To Restore John Deere
Two-Cylinder Tractors**
0-7603-0979-5 • 134861AP

**How To Restore Your Farm Tractor
2nd Edition**
0-7603-1782-8 • 137246AP

**Mustang 5.0
Performance Projects**
0-7603-1545-0 • 137245AP

For a FREE catalog visit **WWW.MOTORBOOKS.COM** or call 800-826-6600